Textile Testing

Textile Testing

Jewel Raul

APH Publishing Corporation
Ansari Road, Darya Ganj, New Delhi-110 002

Published by
S.B. Nangia
A P H Publishing Corporation
4435-36/7, Ansari Road, Daryaganj
New Delhi 110002
Ph.: 23274050
E-mail : aphbooks@gmail.com

2026

Printed at
Balaji Offset
Navin Shahdara, Delhi 110032

DEDICATION

This book is dedicated to GOD, THE HEAVENLY FATHER, as a token of my love for my beloved son ROBERT ROSHAN and his wife RAINE IPSITA wishing them a very happy married life of ever-togetherness.

Acknowledgement

I owe the most, to my Heanenly Father, for strenthening and inspiring me constantly for the hard labour with dedication incurred for equiping this bundle of papers to take the form of a book. I also owe much to the authors, whose books are referred, illustrations cited, and figures quoted.

—Jewel Raul

PREFACE

'Textile' literally means material that is made by uniting yarns or fibres together to form a mat or sheet of any unlimited length or width manufactured by weaving, knitting, felting, bonding etc. having been used for covering, warmth, personal adornment and even to display wealth, from time immemorial.

There are many fibrous structures in nature, but it is only those which can be spun in to yarns suitable for weaving or knitting which can be classified as textile fibres. In order that it may have commercial value, a textile fibre must possess certain fundamental properties. It must be readily obtainable in adequate quantities—at a price which will not make the end-product too costly. It must have sufficient strength elasticity and spinning power. The later property implies a measure of cohesion between individual fibres, which will give strength to the yarn when they are twisted together.

The spinning of fibres, no doubt has helped when there is a certain amount of surface roughness or serration and it is also promoted by fineness and uniformity in diameter. In addition to these fundamental properties, there are other which are desirable such as durability, softness absence of undesirable colour, and an affinity for dyes. Some fibres have few and others have many of these properties. Some of the valuable properties such as colour or softness, may be latent and the object of finishing process is to develop them to the highest possible extent without anyway damaging the fibre.

Textile testing enables to explore this truth which are or may be done in different stages like fibre, yarn and in fabric and the result is revealed by showing the speciality. Thus it is a valuable aid to the textile manufacturers, distributers, retailers and to the most, the consumer.

—Jewel Raul

Contents

1

TEXTILE TESTING

Testing refers to anything used to distinguish or defect substances in any matter.

Textile refers to anything that is capable of being woven or the aspects pertaining to weaving. Thus Textile Testing as a whole refers to the vigorous testing done on textile materials which may be inside the laboratory as well as in its natural setting or in the day-to-day uses, using various testing aids, equipment, simple as well as sophisticated and technical instruments by use of different chemicals and reagents in different physical, physiological and environmental conditions upon various natural biological mineral elements and resources under different reactive and non-reactive situations for the development of textile industry for effective and satisfied textile consumership.

Textile testing is a valuable aid to those engaged in the production, distribution and consumption of textile if the instruments and techniques are used effectively. The results of any text must be studied carefully in order to enable to take the right course of action. In fact testing is a means to an end and not an end in itself. Persons involved in textile testing must be technically qualified Exa. The Textile Technologists. They must be part scientists, part statisticians, part technologists and part diplomats. Simultaneously, they must keep abreast in new developments in processing. Diplomacy is required by the part of the textile technologist to put-over his findings to coleagues at management and in the mill with the workers.

There are varieties of instruments that are used in Textile Testing. Some of them are simple, some important, some are electronic to handle carefully, and some are new involving high technologies. In this case choice of appropriate instrument for particular test is important.

Subject of Testing

This involves so many 'w' s covering questions as to what why, when, who and whom about the tests.

Objective of Testing

The objective of Textile Testing can be varied which are summarised as under:

1. Research: The roads along the research works to travel is cross-road, fork roads, bridges and culs-de-sacs (Street etc closed at one end). There is a choice of direction at each stage. Results of testing in research helps the scientist which route to follow next. What appears to be sound in theory is often disproved by experiment.

2. Selection of Raw Materials: The raw materials for:

Spinner	-	Fibre
Weaver	-	Yarn
Finisher	-	Fabric

One attribute common to most raw materials is their variation in quality. Fibres vary in length colour and fineness. Yarns vary in court, strength and twist. The fabric vary in their thread-per-inch, freedom from faults and shrinkage. Prevent is always better than cure. Thus testing of available raw materials is done to ensure smooth running of production process. In this way unsuitable materials are rejected and put for another use. The standard by which raw materials are rejected or to be rejected must be realistic, otherwise, it will reject much raw materials which infact are good enough. On the other way, large amount of inferior or useless materials will find its way in to the flow of production and cause trouble. This testing is not so important incase of man made fibres or continous filament their properties are controlled during manufacture.

3. Process-Control: Any processing should be supervised or controlled while in operation because when processing goes out of control the amount of waste and the number of seconds increase, cost goes up and temper also goes 'high'. Higher end breakages in spinning and winding departments and excessive loom stops due to warp or weft breaks, affect the operatives as well as production. A plan of production requires certain standard levels to which materials in process must conform. For this, the quality control scheme measures are weight-per-unit length of lap, sliver roving or yarn. For achieving maximum effectiveness, the "Process Control" test should be done close to the processing machinery.

4. Process Development: It is considered as a form of applied research. The experimental works are carried out in research institutes on the actual processing machinery. Investigations and further researches are being carried out for better, cheaper and quicker methods of manipulating fibres and yarns. The success achieved is often measured by the improvement in one or more characteristics of the material delivered after change in machine design or setting. It is important to be quite clear while testing which properties are to be measured to avoid unnecessary waste of time and money.

5. Product Testing: The object of a product test is to assess or findout the performance quality of a finished article in actual service. The assessment of the resistance of a fabric to wear and tear of every day use is the most difficult problem. For this purpose several imitative test like rubbing against abrasives, wetting and drying bending, stretching, creasing and so on are carried out. It is done alternatively subjecting the materials to a series of lab tests each test for only one property at a time.

(vi) Specification Test: Production of textiles to meet specification has growing demand. Its advantages are:

(a) Prevention of deterioration in quality by manufacturers using inferior raw-materials.

(b) Production of goods of known performance.

(c) Opportunity for a manufacturer to produce exactly what is required by the customer.

(d) The customers really know what he wants and can frame a specification in the right way vague specification, due to inability of the customer to explain what he want in precise term leads to more than one interpretation which results in unsuitability of the finished product for the intended purpose. One type of specification test is producing a small sample with a request to "product this for me please" for which suitable analysis of techniques and testing are required plus the need to be aware of the inadequate size of the sample.

Testing Bureau (Tested Quality Schemes)

Testing play an important part in the operation of tested quality schemes. The materials tested here are usually in the finished state that is in the form of fabric. The end-product is of-course the result of all the care or lack of care which has been given to the choice of raw-materials. The processing, the finishing and the making up. There are a number of organisations formed to test the performance of textile some of which are independent bodies to carry-out tests on materials submitted by manufacturers and issue certification marks and allow the manufacture to indicate on their levels that the goods have been tested and found satisfactory. Other schemes are operated whose objective is to help the manufacturers, the retailers and the customers to gain the maximum satisfaction from their textiles.

The tests carried out on the materials and garments are related to their end uses. For exa-swim-suits are tested for colour-fastness to sea-water and chlorinated water and industrial overalls are tested for shrinkage in machine washing. Minimum performance standards have been set up and the articles of fabric is rejected, if the test results fall below this minimum standard. Some bodies test particular aspects of the materials (American Institute of Landering) Exa-some Institute use their own"Certified washable seal" to goods which launder to their standards.

"Uster Analysed" is an interesting trademark which signifies that the yarn sold is manufactured under an efficient quality control-system. Another trademark with wool-mark granantees than the goods are made from wool and possess the desirable attributes of pure wool. A recent organisation "housewife" have begun to test all kinds of goods including textiles to recommend for a "good buy to which one may become member by subscription and receive regular report on the article tested."

2

HUMIDITY-EFFECT OF MOISTURE ON PROPERTIES, STANDARD CONDITIONS AND CONDITIONING

Some of the most important properties of a textile fibre are closely related to it's behaviour in various atmospheric conditions. Most fibres are hygroscopic, that is they are able to absorb water vapour from a most atmosphere and conversely desorb or loose water in a dry atmosphere. Many physical properties of a fibre are affected by the amount of water absorbed, dimensions, tensile strength, elastic recovery, electrical resistance, rigidity and so on. In a fabric, the moistrue relationship of a fibre play a major part in deciding whether the fabric is unsuitable for a particular purpose. The structural details of the fabric can modify the apparent behaviour of the fibre. For exa-fabrics woven from a hydrophobic material such as 'Terylene can pick-up water by a "wicking"action along the fibre and yarn surfaces.

Regain and Moistrue Count:

The amount of moisture in a sample of material may be expressed in terms of Regain or Moisture Content. Regain is defined as the weight of water in a material expressed as a percentage of the oven day weight. Moisture content is the weight of water in a material expressed as a percentage of the total weight.

Let oven dry weight = D
Weight of Water = W
Regain = R
Moisture content = M

Then R= 100W/D and M $= \frac{100W}{D+W}$

Also R = M/1-(M/100) and M $= \frac{R}{1+CR/100}$

Atmospheric Conditions and Relative Humidity

Along other things the regain of a textile material depends upon the amount of moisture present in the surrounding air. The dampness of the atmosphere can be described in terms of "humidity" either absolute humidity or relative humidity.

Absolute humidity: Absolute humidity refer to the weight of water present in unit volume of moist-air i.e., grains per cubic foot or grams per cubic metre.

Relative Humidity: Relative Humidity is the ratio of actual vapour pressure to the saturated vapour pressure at the same temperature, expressed as a percentage.

$$r.h.= \frac{\textbf{Actual vapour pressure}}{\textbf{Saturated vapour pressure}} \times 100(\textit{per cent})$$

Moisture Determination as related to textile

An alternative definition for relative humidity is the ratio of the absolute humidity of the air to that of air saturated with water vapour at the same temperature and pressure. This ratio may then be expressed as a percentage. It is convenient to describe a given atmosphere in terms of relative humidity rather than absolute humidity because the regain of textile materials appears to depend upon the relative humidity rather than the actual amount of water vapour present. Since the relative humidity affects the regain of a textile material and since the properties of the material are influenced by the regain, it is necessary to specify the atmospheric conditions in which testing should be carried out. The instruments used in determining the humidity are shown as hygrometers or psychlrometers. There are other methods such as the gravimetric, chemical and the dew-point methods, but they are not commonly used. The three main types of instruments used in this purposes are:

(i) Wet and dry bulb hygrometer.

(ii) Hair hygrometer.

(iii) Electrolytic hygrometer.

Effect of moisture on properties

The following are some factors affecting the regain of textile materials:

(a) Time: A material placed in a given atmosphere takes a certain testing. For amount of time to reach equilibrium. The "rate of conditioning" depends on several factors, such as the size and form of the sample, the material, external conditions etc. This time element is reflected in the procedure for testing. For example in testing yarn for count, it is recommended that" prior to reeling conditions, the samples in the atmosphere for testing is kept for not less than one hour for yarn in hand and for not less than three hours for yarn in all other types of package, and reel them in that atmosphere.

(b) Temperature: For practical purposes, the effect of temperature on regain is not important, but it is the relative humidity which plays the major role. A change of 10°c will give a change in the regain of cotton of about 0.3 % but since it is likely that the testing room temperature will ever be 10° or 30° C, the effect can be ignored.

(c) The Previous history of the sample: The histories effect is a good example of the previous history of the sample effecting the equilibrium regain. Processing can also change the regain. When oils, waxes, and other impurities are removed, the regain may change. Example: The official regain for scoured wool is 16% and for oil-combed tops it is 19%. Therefore, in the study of regain values, the physical and chemical history of the material must be taken into account.

The Common Properties which are Affected by the Moisture in Atmosphere are :

(1) Dimensions: Absorption of moisture is accompanied by changes in the dimension of fibres. Swelling is mostly transversal since the water molecules penetrate between the more or less parallel molecular chains and exert their forces outward. Dimensional changes in the fabric as it passes from one atmosphere to another may result in the wrinkled appearance suits, tailored in one atmosphere and worn in other especially in climates where high humidity are found. This advantage of swelling is taken in to consideration in the design of closer as the component yarns swell, thus, preventing the penetration of water.

(2) Mechanical properties: The general effect of water molecules in the fibres to reduce the size of the forces holding the molecular chains together thereby weakening the fibre. (Important exceptions to this are found when the vegetable fibres such as cotton and flax are considered an increase in strength is noted). In this the maximum strength not only is reduced but the stress-strain curve also assumes a different form. The yield-point may be shifted, and applied loads which would not cause any damage on dry fibres, but it may stress the fibre beyond the lowered yield point. Other mechanical properties affected by regain include extensibility, crease-recovery, flexibility and ability to be set by finishing processes.

(3) Electrical properties: The "before and after" effect of moisture on the electrical resistance of textile material is most striking. The ratio of the resistance at low regain and at high regain can be at the order of hundreds of thousands to one, a circumstance which has led to the design of moisture-meters based on the measurement of resistance value of textiles. Other electrical properties affected by the amount of moisture in the material are the dielectric characteristics which are a source of effort in the measurement of the irregularity of slivers, rovings and yarns on capacity type testers.

(4) Thermal Effect: When moisture is absorbed by

textile materials, heat is generated. The heat is referred to as "heat of absorption". Suppose the dry-weight of a sample is 1 gm and it is then completely wetted, the heat evolved expressed in calories per gram of dry material is termed the "heat of wetting". A practical example of the importance of this effect is seen when clothing is considered. In winter a person passing from indoors to outdoors usually goes from a warm room with a low percentage r.h. Relative Humidity to a cold environment with a higher percentage r.h. The regain particularly of the outer garments increases and heat is generated, so acting as a buffer to the sudden shock of a change in the temperature which the body would otherwise suffer.

Moisture-Regain and Moisture-Content

The methods employed in the determination of regain vary in refinement and accuracy and the method chosen may be directly related to the object of the test. In research work great accuracy is usually demanded and period of several months may be required to complete the measurement. Regain is the weight of moisture in a material expressed as a percentage of the oven dry-weight. It follows that the basic method of measuring regain must be to weigh the sample in its original conditions to dry it and then to weigh it again. The 'oven-dry-weight' is defined as 'the constant weight obtained by drying at a temperature of 105± 3°C.

In 1964, the drying condition specified include a recommendation to use a ventilated drying over with a positively induced air-current. When successive at intervals of 20 min., differ by less than 0.05%, it may be assumed that a constant weight has been reached. The adequate sampling scheme should always be used where possible so that the samples on which the tests are made are the representative of the bulk. Sometimes, the sample contains oils, sizes and other types of added substances which should be removed before oven drying.

3

FIBRE-TESTING

A single fibre is the unit from which many complicated textile structures are assembled. It is a very small beam characterized by great length relative to it's cross-section. This unit is itself a complex-structure built from atoms and molecules.

The two important dimensions of fibre are length and fineness and there is sometimes a relationship between these two qualities, when natural fibres are considered. In case of man-made fibres both the length and the fineness can be controlled independently. Of course, there is influence of natural fibres when the man-made fibres are to be processed on machines, designed to suit the dimensions of natural fibres. Further when natural and man-made fibres are blended, their dimensions may be compatible. The measurement of the natural fibres is a difficult task because there is variation in between types of the same material as well as the same type.

Exa. : The sea-island cottons are longer than American cotton but within a sample of sea-island or American cotton, there is variation in length between fibres. For this reason, it is necessary to derive a single length value to characterize a given sample and to obtain an index of the variation present.

Fibre-Length-Measurement

In a length distribution, the total length of fibres in each class is calculate. Thus, the fibre-length in each class is the class length muliplied by the class-frequency.

Cotton is grown in different parts of the world. After-picking and ginning, the bales have to be assessed for quality. One of the attributes of cotton which requires measurement is the "staple-length". The definition of staple-length is— "a quantity estimated by personal judgement by which a sample of fibrous raw-materials characterize as regards as technically most important fibre length". The determination of length of cotton consists of selecting a sample and preparing the fibres by hand-doubling and drawing to give a fairly-well-straightened tuft about half inch wide. This is laid on a flat-black-background and the staple length measured. The two methods of determination of wool length characteristics are the single fibre and the tuft methods. Single-fibre methods have particular merit for research purpose or for cases, where the time element is of minor importance.

The term" wool fibre length" is really an indeterminate expression somewhat akin to the "length of a piece of elastic". A wool fibre not a naturally wavy configuration when free from applied tension and the crimp is multi-planar, i.e., the kinks just out in all directions. If a fibre were gripped at one end and an increasing force applied at the other, it would gradually straighten-out until the crimp was removed. It's length at any instant would depend on the load at the moment. It is therefore necessary to define the following terms used in connection with wool.

(a) Crimped length: The distance separating the ends of the fibre when it is not constrained.

(b) Fibre length: The estimate of the distance between the fibre-end when a tension just sufficient to remove the crimp has been applied.

Even the later definition is not too precise since different operators may have different conceptions of the point when the crimp is just removed.

Flax: The length for flax-fibre is determined by a sorting machine designed by the Linen Industry Research Association.

Asbestos: The length properties of asbestor fibres may be assessed with the aid of a comb sorter using appropriate technique.

Measurement of Fibre Fineness

The importance of fiber fineness:

In the determination of the merit of a material for spinning, the fibre length is often taken as the criterion. Besides for many other purposes, the fineness of the fibre is of equal and sometimes of greater importance. For a given count the average number of fibres in the cross-section will depend on the fibre-fineness. The finer the fibre, the higher number, and the lower irregularity. Starting with a coarse fibre, the spinning limit is reached fiarly soon. Summarising these idea, two points emerge.

(1) If a given count is spun-from a fine and coarse-fibre, a more uniform and stronger yarn will result from the fine fibre.

(2) A fine-fibre can be spun to finer counts than a coarse-fiber. Broadly speaking, the finer the fibre, the greater the total surface area available for inter-fibre content and consequently less-twist is needed to provide the necessary cohesion. This is reflected in the twist factors used for different types of material. The fineness of the fibre also affects several mechnical properties and therefore influence the behaviour of the fibre during processing and the properties of the resultant yarns and fabrics. The following are those two important properties.

(a) The torsonal rigidity or resistance to twisting.

(b) The stiffness or resistance to bending.

Definition of fineness:

If all fibress had perfectly circular cross sections it would be a relatively simple matter to measure the diameter by means of a microscope. The task would be even simpler if it were-known that the cross-section was uniform along the fibrelength and from fibre to fibre. Since, the fibres exhibit a variety of cross-sectional shapes and they also vary in section along their length and vary from fibre to fibre. Therefore, it is necessary to derive some index of fineness which can overcome these difficulties and at the sametime, be comparatively easy to determine.

The mass of cylinder or prism is given by—

Mass = Volume × density

= Cross-sectional area x length x density.

From this we see that, mass is directly proportional to the cross-section. Further, the mass of a known length of material will be directly related to the cross-section. By measuring the weight of a known length of fibre, we obtain quantity called the linear-density-which can be expressed in terms of weight per unit length. It should be noted that, it is quite possible to have fibres with identical linear density but different cross sectional areas. For example, a fibre with a high density will have a smallar cross sectional area than a fibre of low density.

Principle of Fineness-Measurement

Gravimetric Method

In the gravimetric method, the weight of the mean fibre per centemeter is calculated. This result may be corrected to give the standard fibre weight per centimeter after maturity of the cotton has been determined. The optical method includes the technique of micrometry in which it is important to remember that the shapes and diameters of the natural fibres on cross-sections are subject to a greater degree of variation than those of man-made fibre. To obtain a reliable result, it is threfore, necessary to measure a large-number of fibres. The more variable the measure dimension, the higher the number will have to be to give result to a desired accurary. In fibre-microscopy, cross-sectional area is measured by projecting onto graph-paper with 10 or more images of reactions. The scale is caliberated in microns by means of a stage-micrometer and the number of squares covered by the fibre-sections counted.

The profile method for wool is done by using a projection microscope to measure the fineness of wool. This is done by mounting sample of very-short out length of fibre on a suitable mountant and then project their images on to a screen carrying a movable transparent scale. The length of the pieces are 1mm or less and under the microscope they look like little cylinders. The choice of mountant is important since it must not absorb moisture and it must have a refractive index of between 1.53 and 1.43 (to give a well-defined image), and it must be capable of dispersing the tiny cylinders of wool. Cedar wood oil and liquid paraffin have been found suitable. Since the accuracy of estimator of fibre-diameter by projection micro-scope, is not of a high order when the fibre has an irreqularly shaped cross-sction. Measurement of fineness is also done by an equipment called Vibroscope. It ditermines the fineness by recording the vibration as the finer the fibre the more vibration will it have.

Exa - Strong of violin E, A, D, G, respectively the E String is the finest with highest pitch, G, the coarest with low pitch. From this preamble, it may be inferred that frequency, tension and fineness are related quantities.

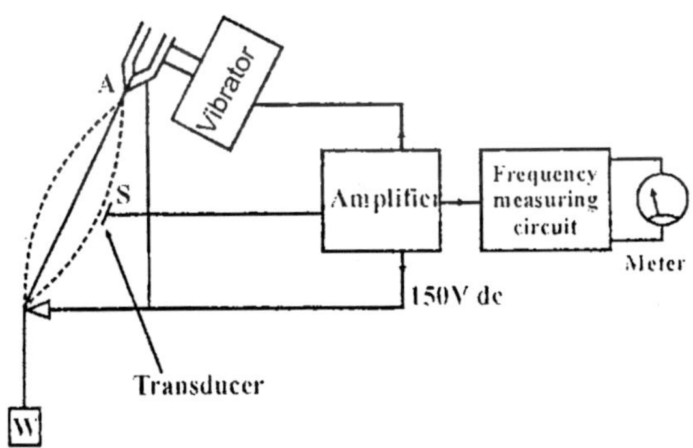

Fig. 1 (a) : The B.R.R.A. Vibroscope.

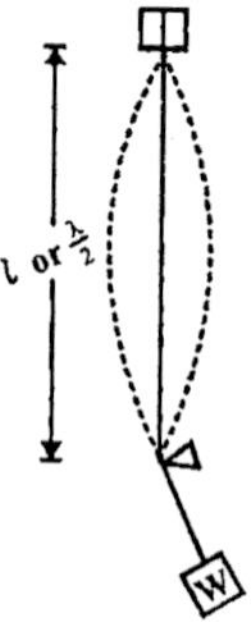

Fig. 1 (b)

Air-Flow-Method

In this, a sample of a known weight is compressed in a cylinder to a known volume and subjected to an air current at a known pressure. The rate of air that flow through this porous plug of fibre is measured. The flow meter often being caliberated in terms of fineness instead of volume per unit time, e.g. microns for wool, microgroms per inch for cotton. In this given diagram, the rate of air-flow through (b) is less than that flow through (a), even though the space through which the air passes is the same for both cyliders. Hence difference in the rate of air-flow is the measure of the difference in the surface area of the large diameter and small diameter rods.

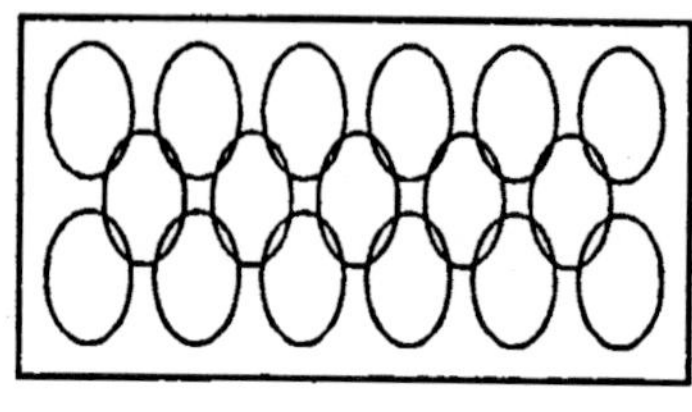

Fig. 2(a)

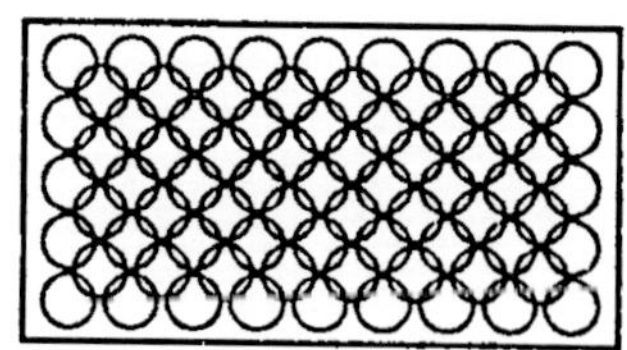

Fig. 2(b)

MATURITY OF FIBER

The maturity ratio-concerns, the development of the cell- wall. The perimeter of the cotton fibre is least affected by environment and most by heredity, but cell-wall thickening is highly sensitive to growing conditions. Even when cotton is grown under favourable conditions a small proportion of the fibre will be under-developed or immature, but adverse weather, poor soil, plant disease and pests etc., will increase the proportion of immature fibre and lead to trouble in processing.

Immaturity in tufts tends to remain in patches and immature fibres tend to associate even through to the yarn stage. Small-patches of crop can be affected and the chances of maturity variation are great in any sample or bale of cotton. One of the main trouble caused by the presence of these thin walled immature fibres is nepping, which is created during processing, starting at the gin, apart from the natural causes like fragments of seed pod attached to a fibre. Immaturity also affects the shade after dyeing. As a result, there is difference in the shade of the colour on the yarn. Fine cotton tends to be lighter in shade than coarse cotton.

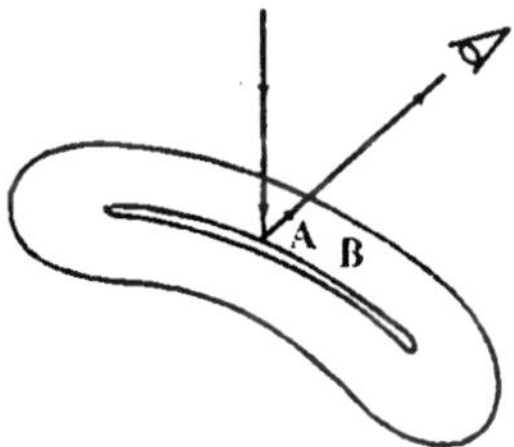

Fig. 3(a) : Reflection of light from the internal surface of a cotton fibre.

When the colour component from the dyed fibre reaches the eye, it is mainly composed of rays of light replected from the internal surfaces of the fibre. It can be observed in fig. 3(a) that, the light reflected from point A will be the colour of the dye in B. The remnant of the incident light which has not been absorbed by the fibre is the colour that we see. The greater the depth of B, the deeper the shaal.

In order to assess the quality of cotton with respect of maturity and its effect on ends down, yarn strength, dyeing troubles, and so on, some methods of measurement is required. The degree of cell wall thickening may be expressed as the ratio of the actual cross-sctional area of the wall to the area of the circle with the same perimeter as seen in fig 3(b). In this test, the weight of the fibres are determined and 5 tufts of cotton are taken. Each tufts laid on the microscope slide, parallel but separated and a cover-slip put

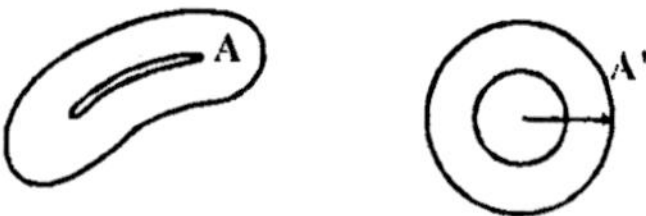

Fig. 3(b) (Degree of thickening of a cotten fibre)

over the middle. Then the fibres are irrigated with a small amout of 18% caustic soda solution which has the effect of swelling them. The presence or absence of convolutions is then observed, preferably by means of a projection microscope. This enables the fibres to be classified in to three groups:

(1) Normal fibres,
(2) Thin-walled fibres,
(3) Dead-fibres.

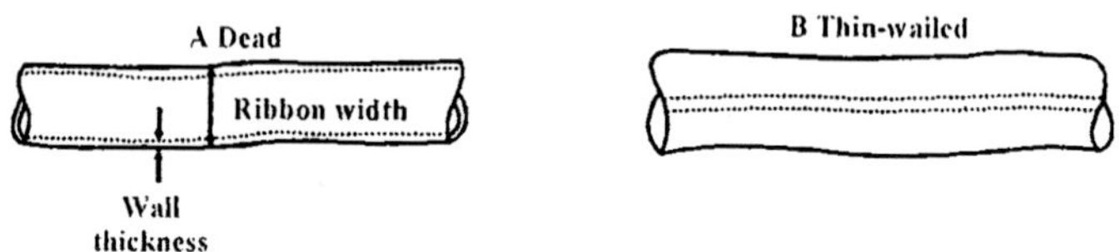

Fig. 3 (c) : 'Dead' and 'thin-walled' cotton fibres after irrigation with 18 per cent caustic soda. Fibre a is classed as 'dead' because the wall thickness is less than one-fifth the ribbon width.

Mature fibres with a well developed cell wall and pronounced convolutions in the raw-state become rod-like after swelling. These rod-like fibres are classed as 'normal'. Dead fibres appear ribbon-like even after swelling. Thinwalled fibres are those lying between the other two classes. Borderline cases may be judged from the thikness of the cell wall, relative to the total width of the ribbon. If the wall is less than one fifth of the total width, the fibre is calssed as dead.

Test Maturity—Differential Dyeing

Another test for maturity is the Goldthwaite test. Faulty dyeing is a trouble that occurs in case of immature fibre. This test gives a visual indication of the maturity of a sample of cotton. In this method, two dyes are used in the same dye both, one red and the other green after dye bath, the mature fibres are stained red and the immature fibres turns green. It is because, the red colour which is used in the dye-bath being developed in the cellulage of the secon-dary wall. Hence little or no secondary wall thickening in the fibre no red.

Testing Maturity by use of Polarised Light

Use of polarized light is another method for of deternmining maturity fibrwhare. The appearance of fibre maturity is observed under a microscope fitted with a suitable polarising equipment. The result shows some colour, where fibres are clossed as immature, partially mature or mature. With a sample of fifty fibres total number coming under the groups gives the percentage of naturity.

Fibre Quality

The important characteristics of cotton which are included white assessment of its quality are fineness, length, maturity, unifornity and grade. Apart from the other, the quality of 'grade' is a term which needs to be explained. The spinner of yarn requires a certain degree of uniformity and continuity in the quality of the raw materials delivered to his melt, when the type of cotton has been decided upon. It is a universally known factor that, cotton quality varies from year to year, field to field and bale to bale. Thus, to avoid the problems of continual re-adjustment and settings of the processing machinery to be done frequently, some system of classification or grading must be used. This ensure that the spinners receives their cotton in smoothly running late. Gradings also face variation in the methods used in grading, differences in terminalogy, difference between centres separated by great distances like American, Egyption, Sudan etc. So the different grading systems are also taken in to consideration. The graphs shown in Fig. 4 shows the effect of a length assessment and the reduction of the importance of trash content of cotton from Port Sudan and Gezira Plain as compared with American grading Practice.

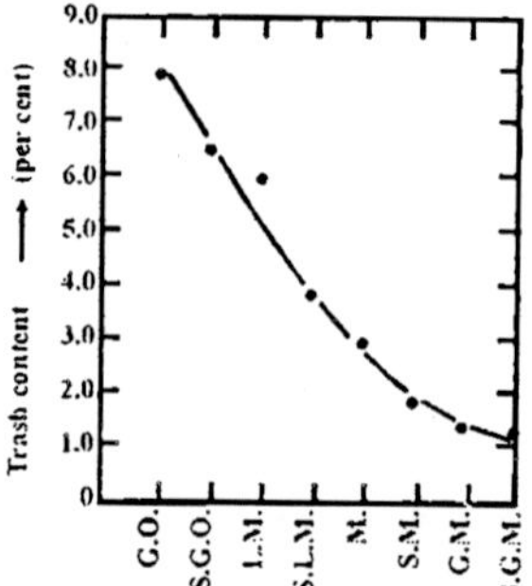

Fig. 4(a) : Analyser trash content and American grade.

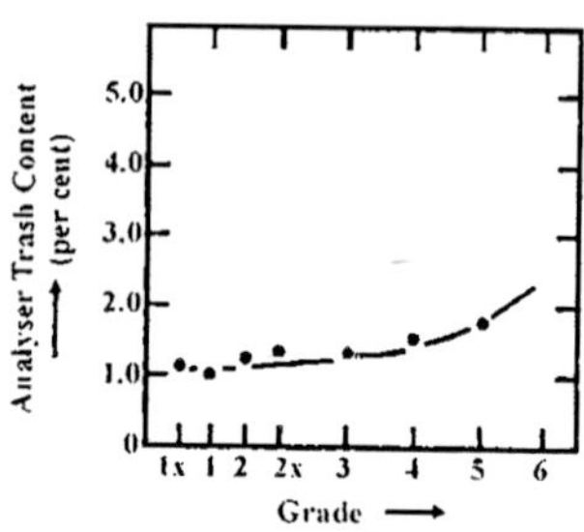

Fig. 4(b) : Analyser trash content and Sudan-Egyptian grading.

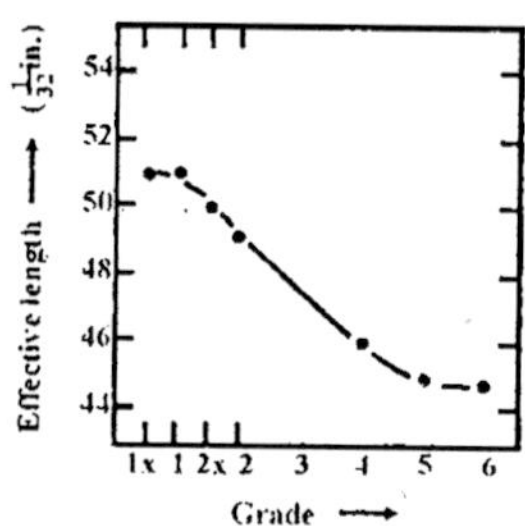

Fig. 4 (c) Domains Sakel grown in Gezira. Effective length and Sudan-Egyptian grading.

4

YARN TESTING

The yarn dimensions and structural details are dealt with the count and twist of the yarn. The count of a yarn is a numerical expression which defines it's fineness. Spun yarns are only, roughly circular in cross-section. And irregularity in thickness is unavoidable. Filament yarns with only a small amount of twist in them are referred to as "flat" yarns, possibly because they flatten so easily when in contact with other more solid bodies. A definition of yarn count given by Textile Institute is "Count—A number indicating the mass per unit length or the length per unit mass of yarn.

Note- The various counting unit systems using different units of mass and length are in use, so the system used must be stated".

Direct and indirect system of yarn numbering

Mass per unit length—Direct.

Length per unit mass—Indirect.

Direct systems: In a direct yarn counting system the yarn number or count is the weight of a unit length of yarn. The units of weights and length vary from trade to trade and district to district, a state of affairs which results in a multiplicity of counting system.

Let N = The yarn number or count.

W = The weight of the sample at the official

regain in the units of the system.

L = The length of the sample.
l = The unit of length of the system.

Then $$N = \frac{w \times l}{\ell}$$

Example: If a skein of 100m of filament viscose yarn weight 1.67 calculate its denier.

In the denier system, the weight unit is the gram and the "unit of length is 9000 m. The denier of a yarn is the weight in grams of 9,000 m.

Thus W= 1.67 g, L= 100 m and 1= 9,000m..

Therefore. $\text{Denier} = \frac{w \times l}{L} = \frac{1.67 \times 9{,}000}{100} = 150.3$ Denier.

Indirect system: In an indirect system, the yarn number or count is the number of units of length "per units of weight" Here again there are various units of length and weight and numerous systems. Generalising:

Let N = The yarns number or count.
W = The weight of the sample at the official regain kin the units of the system.
w = The unit of weight of the system.
L = The length of the sample and
1 = The "unit of length" of the system.

Then $$N = \frac{L \times w}{l \times W}$$

Example: A lea (120 yd) of cotton yarn weighs 25 gr. calculate its count in the cotton system.

In this case the "unit of length" is the hand, 840 yards, and the unit of weight is 1 lb. In 1 lb there are 7,000 gr. Hence, L=120 yd, l= 840 yd, W = 25/7,000 lb and W = 1lb

Therefore: $$\frac{120 \times 1}{840 \times 25/7000} = \frac{120 \times 7000}{840 \times 25} = 40s$$

(The small 's' after the count number is convention used for expressing count.)

Note: The values 7,000 and 840 appear in all such cotton yarn calculation and may therefore, be reduced to a constant fraction 100/12 i.e., 7,000/840. A useful little formula is then evident:

Cotton count = (Length in yards/weight in grains) x (100/12).

(The small 's' after the count number is a convention used for expressing count).

Note: The values 7,000 and 840 appear in all such cotton yarn calculations and may therefore be reduced to a constant fraction 100/12 i.e., 7,000/840. A useful little formula is then evident:

$$\text{Cotton count} = \frac{\text{lengh in yards}}{\text{weight in graing}} \times \frac{100}{12}$$

Further, if leas are always tested then:

$$\text{Cotton count } \frac{120 \times 100}{\text{Weight in grains} \times 12}$$

$$\text{or } \frac{\mathbf{1000}}{\textbf{Weight in gai}}$$

Counting Systems:

The textile industry grew-up in different localities and each chose its own system to suit local conditions. Conversion from one system to another is achieved by conversion factors and constants. A multiplying conversion factor is used where a direct to direct system or an indirect to indirect system is concerned. Where the conversion is from direct to an indirect or vice-versa, a constant is used into which the known count is divided to give the equivalent count in the other system. The yarn number or count on 'tex-system is the wieght in grams of 1 km of yarn. This system is therefore, a direct system, simply defined and simple to use. The extensions of the basic system are used in order to express the fineness of fibre and filament on the one hand and intermediate products and coarse yarn on the other.

For example- For the fibres, the fineness is expressed

in "millitex" i.e., the weight in milligram per kelometer and for an intermediary products such as slivers and course yarn structures like cords, the count is expressed in "kilotex" i.e., the weight in kilograms per kilometre or what amount to the same things, grams per mitre.

Length Measurement

The yarns are wound in to reels or bobbins by a simple machine consisting of reels, yarns package creel, a yarn guide taking small sideway traverse to spread the loops of yarn, a length indicator and a warning boll. The wrap reel marks that 80 revolutions of the reel produces skein of 120 yd, or a lea. The limits are set on the skeins gauze scale with a pair of movable pointers.

Weight Measurement

Balances—The analytical balances and any other special yarn balances used in the determination of count must be accurate and it is necessary that they are well-maintained and that before use they are levelled and checked.

Regain—The problem of accounting for the presence of moisture in the sample can be tackled in several ways two of which are considered here.

(i) Determine the over-drey weight and multiply by 100 plus standard regain devided by 100.

This means — oven-dry weight $\times \dfrac{100 + \text{standard regain}}{100}$

(ii) Allow the sample to conditions in the testing atmosphere long enough to reach equilibrium and then weigh in the same atmosphre.

Count Calculations

Yarns prduced by combination of two or more single yarns are used in many textile structures in order to achieve effects which may not be readily obtained if single yarns are used. It is often necessary to dissect the sample and carry out count tests on short lengths, where the count of the thread must be meausred. In an indirect system of yarn counting, the reciprocal of the resultant yarn count is equal to the sum of the reciprocals of the component treads neglecting the effects of twisting on length.

N = resultant count

N^1 N^2 = component thread counts

Thus $$\frac{1}{N} = \frac{1}{N_1} + \frac{1}{N_2}$$

Inc our example, N^1 and N^2 are known to be the same

Hence $$\frac{1}{N} = \frac{1}{N_1} + \frac{1}{N_2} = \frac{2}{N_1} = \frac{1}{14}$$

$$N_1 = 28s$$

Now, because the contracton has, in effect, caused the single yarns to become coarse, we must correct our estimate by making it 10% finer, and so the corrected estimate of the original singles count will be 28 x 1.1 i.e., 30.8s. Sometimes doubling of yarns causes the component threads to increase in length and the correction to the estimate will go the other way.

Example: A fairly simple yarn and the other with plied structure is built by doubling yarn which have themselves been doubled.

Yarn Count and Yarn Diameter

The geometry of fabric structue is an increasingly important branch of textile technology. It is a subject which forms a rational foundation upon which a scientific study of fabric construction is based. It deals with the appearance handle, drape and general behaviour of fabric which are dependent upon many factors such as materials used, yarn structure, weave and finish. One important dimension in the study of fabric geometer is the 'diameter' of a yarn, that is the circular cross-section of yarn.

A number of simple formula are available for estimation yarn diameter :

cotton yarn

(i) Diameter (in inches) = $\frac{1}{\sqrt{(\text{yards per pound}) - 10 \text{ per cent}}}$

(ii) Diameter (in inches) = $\frac{1}{26.1\sqrt{\text{counts}}}$

(iii) Diameter (in inches)= $\frac{1}{\sqrt{(800 \times \text{counts})}}$

worsted yarns

(i) Diameter (in inches) = $\frac{1}{\sqrt{(\text{yards per pound}) - 10 \text{ per cent}}}$

(ii) Diameter (in inches) = $\frac{1}{21.3\sqrt{(\text{worsted Counts})}}$

(iii) Diameter (in inches) = $\frac{1}{\sqrt{(500 \times \text{worsted counts})}}$

Yarn Twist and Twist Measurement

The word "spinning" embraces a number of processes that can transform fibrous raw materials into yarns. This word spinning is also used in case of man-made continous filaments to make them solid and firm yarn. In a more clear explanation, a strand of fibre is given a number of twist on it's axis to form a yarn. "Twist is the measure of the spiral truns given to a yarn in order to hold the constituent fibres or threads together (Skinkle). Morton said when a strand is twisted, the component fibres tend to take on a spiral formation, the geometric perfection of which depends on their original formation. In some other definitions. Twist is the rotation around the yarn axis of any line drawn on the yarn which was originally, i.e., before twisting parallel to the yarn axis. The Federal Textile Institute defines "Twist"— The spiral disposition of the components of a thread which is usually the result of relative rotation of the two ends.

As twist gives coherenc and strength to a yarn, it is important to consider how much twist is required and what effects have the varying amounts of twist on the yarn properties? The direction of twist at each stage of manufacturing is indicated by the use of the letter S and Z. A single yarn has S twist if the fibre inclined to the axis of the yarn conform in direction of slope to the central portion of the letter S, when it is held in vertical position. Similarly the yarn has Z twist of the fibre inclined to the axis of the

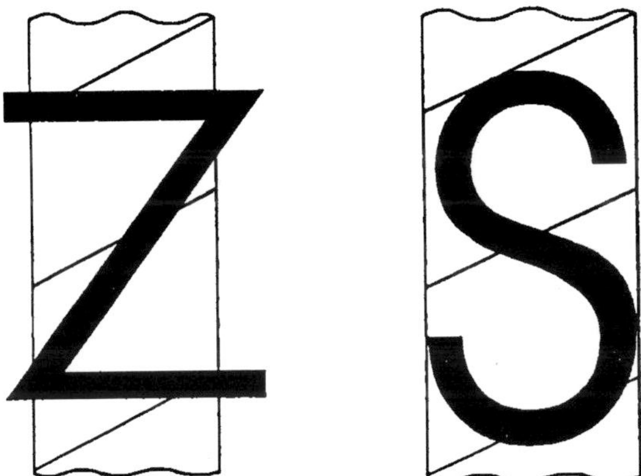

Fig. (3) : Direction of twist in yarns.

yarn confirm in direction of slope to the central portion of the letter Z.

Amount of Twist

The amount of twist in a thread of each stage of manufacture is denoted by a figure giving the number of turns of twist per unit length in the twisted condition at that stage.

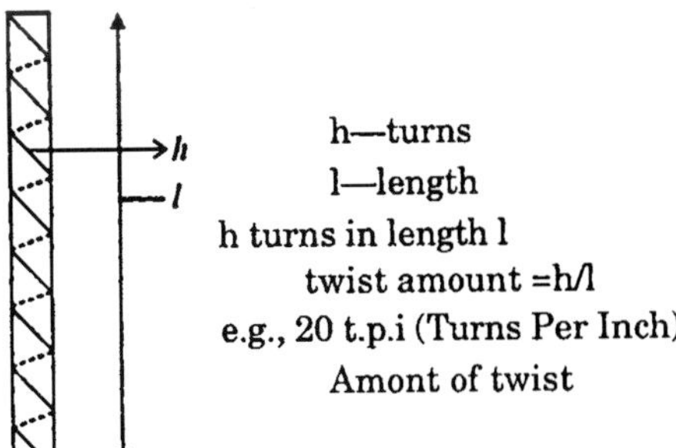

Fig.. 4

A course yarns with 20 t.p.i.has vastly different twist characteristics to a fine yarn with 20 t.p.i. By an expression known as the "twist factor" or "twist-multiplier", it is possible to appreciate the twist character of a yarn even without knowledege of the yarn count.

The Tex System states—Tex twist factor-turns per meter $\times \sqrt{(N)}$.

This results in twist factors of about 2,000 for self twisted yarns and 10,000 for hard-twisted yarns.

Function of Twist in Yarn Structure

Without a twist, a strand of fibres has very little strength to withstand the stresses of preparation and fabric munufactures. It is important to note hère that in some instances, the twistless yarns have been employed as weft in fabric which have shown reasonable weft-way strength. This shows that the yarn properties when in fabric form are not necessarily those demanded by the methods used in preparation and manufacturing processes.

Twist and Yarn Strength

It has been known for many years that an increase in the amount of twist produces an increase in the yarn strength and that this effect holds only up to a certain point beyond which further increase yarn in twist causes the to become weaker.

Some Effects of Twist on Fabric Properties

By varying the amount and direction of twist the fabric designer can achieve a variety of fabric effects. Some of these are visual and some are concerned with handle and drope and some are mechanical, e.g. related to strength or resistance to abrasion.

One example of a usual effect is the 'shadow stripe' if a cloth is woven with the warp threads in alternate bands of S and Z twist. A subdued strip effect is observ5ed in the finished cloth due to the difference in the way the incident light is reflected from the two sets of yarns. In a way, this effect is similar to that, seen when a lown has been mowed and rolled. The twill-line in fabrics based on the twill weave can be subduce or brought into greater prominance by choice of twist direction.

'Shadow stripe' is one example of a visual effect. When a cloth is women with the warp threads in atternate bans of S and Z twist, a subdivided stripe effect is observed in the finished cloth due to the difference in the way, as the incident light is reflected from the two sets of yarns. This effect is similar to the one which is seen when a lawn has been mowed and rolled. The twell line in fabrics based on twill weane can be subdned or brought in to greater prominence by choosing the appropriate direction of the twist. This can be done if.

Example: If the warp of a twill has Z twist and the twill line runs "dwon to the left", then the use of an S-way weft subdues the twill line. Conversely the use of a Z-way weft will product a bolder twill line.

A twisted yarn tends to untwist or assume a configrua-tion in which a state of equilibrium is attained. Highly twisted yarn is "lively" and tends to twist-upon itself and produce 'snarls'. Fabrics made from a highly twisted yarns will possess a lively handle. Crepe yarns for instance, have high twist factors and are used to obtain the characteristic crepe surface. The cloth is woven and afterwards given a wet treatment. Drying is done with the cloth free-from tension. These conditions allow the crepe yarn to curl-up and relax. These Shrinkage occurs and the well known crepe surface is produced. The tendency for yarn to untwist can cause the fabric to curl, especially at the corner. Curling will occur if the in twisting couples of the warp and weft yarns reinforce each other instead of counteracting each other.

Crinp Rigidity Test

The ability of the stretch and bulk yarn to collapse longitudinally is of importance, because the degree of crimp rigidity influences the properties of the fabric produced from these yarns. Crimp rigidity is a measuer of the ability of a textured yarn to recover from stretch and is related to the bulking potential of the yarn.

In a tester developed by the hosiery and Applied Trades Research Association, a load equivalent to 0.1 g per denier is suspended from a skein of yarn which is then immersed in water at room-temperature. After 2 minutes it's length L is measured. The load is then reduced to 0.002 g per denier and after a further 2 min the reduced length L_2, is measured. The cirmp rigidity given by the formula:

$$\text{Crimp rigidity} = \frac{L_1 - L_2}{L_1} \times 100\%$$

It has been suggested that the crimp rigidity test does not give a satisfactory indication of the potential behavior of the yarn when in fabric form. In mill-pracitce, however the test has proved valuable in reducing faults in finishyed fabrics by ensuring that yarns with similar crimp rigidity value are knitted together in to the same fabric. Yarns from different suppliers can be matched for crimp rigidity value and used together.

Twist Measurement

Twist testing methods depends on the specificatio, accuracy, demanded, form of the sample etc. There are a number of points that must be watched for getting reliable result. They are—

(1) Usually twist is not distributed uneformely along a yarn. It has been found that a relation exists between the the twist and the thickness of the yarn at the point where the turns per unit length are measured. This form is "twist × weight = constant". And for to avoiding having bias in twist measurement, twist should be determined at fixed intervals, along the yarn, that is 1 yard of distance which is suitable.

(ii) Effect of withdrawing the yarn over and from the package is another point to consider. Development of twist depends on the yarn type and the direction of twist like S on Z. Yarn can be withdrawn either over the end of the package or from the side. The beter method leaves the twist unchanged.

(iii) A logical method of measuring the twist is to reverse the process and count how many turns are required to untwist the fibres until they are again parallel. It is because twist is inserted in to an element of yarn by the relative rotation of its two ends. This method is known as the straightened fibre method of twist determination.

(iv) Tests on consecative lengths of yarn are not easily made because of the instrument design of some machings and the amount of yarn handling involved. Therefore continuous twist testing is desirable as it has extra advantage of allowing twist tests at fixed intervals without the need for excessive yarn handling which might disturb the twist.

(v) Putting twist in to a strand of fibres on filaments canses the strand to contract in length. On remaining the twist completely from the strand definetely will have longer strand then that of the twisted one. Further twisting of the same untwisted yarn will have the original length.

(vi) A ‘twist to break’ test is the one in which twesting is continued until the yarn breake and the number of twist noted.

(vii) Twist measurement by microscope is done by a microscope having a rotating stage with it’s periphary graduated in degrees. It can be used to measure the twist angle and the yarn diameter and enables the turns per-inch to be derived.

(viii) Twist measurement of plied yarns is to be conducted to deal with measurement of twist and the changes in length observed in each for subsequent twists as plied yarn structure envolves complicated contructions like direction of twist and number of twists.

(ix) The term “Take-up” is defined as the difference in length between the twisted and untwisted thread expressed as percentage of untwisted length.

(x) A quadrant twist tester helps to measure the

doubling twist and the changes in the specimen length at the same time. A step is used which testing single yarns to restrict the pointer movement.

(xi) Twist in component yarn like cabled yarn are different as it is compossed of single yarn twisted to plied and plied yarn twisted to cabled yarn on this, the analysis involves a twist test of cabling twins, a test for doubling twins and finally a test of single yarn twist.

(xii) As per the Testile Standard, 'Designation of structure of yarns are to be given for specificing the structure of all yarns produced by twisting processes in terms of linear density number of plies (or filaments), direction and amount of twists constituent fibres and any other features necessary for a completed description of a yarn.'

5

FABRIC TESTING

Length of the Fabric

Measurement of cloth dimensions should be made in a standard testing atmosphere whenever possible, since humidity may affect the results. It is not always possible to condition the full piece of cloth. The cloth is therefore, measured without conditioning and a correction made by length measurements on a sample.

Width of the Fabric

The width of the fabric when it is removed from the loom may not be the same when it finally reaches the customer. The loom-state fabric may undergo or deals by water, heat, pressure, tension, acids and alkalis. Some processes cause contraction in width (wet-treatment) and some may stretch the cloth (stentering). The textile materials possesses power of recoery from imposed strain and when allowed to relax-free from tension, contraction occurs.

The choice of cloth-width is influenced by a number of factors. Important among this is the end-use of the cloth. For example, handkerchief cloth may be woven so that either two gentle-man's or three-ladies handkerchief occupy the yarn-space in the reed of the loom. Again bedding for single beds will be wov en in narrower looms than those used for weaving for doubled bed fabrics. The maker-up of fabrics buys his cloth in widths which allow him to cut-out his patterns with a minimum of waste.

In the standard method of measurement of fabric width, it is recommended that the fabric should be exposed to a standard atmosphere for at least 24 hours. Before final measurements are taken. Measurements made before and after conditioning will then show whether the change in width, if any, is within the order of accuracy required. On a piece of cloth, ten measurements should be made at points distributed at roughly equal distance trhoughout the full-length of not less than 1 yd should be used and it's width measured at three places. The mean width and the range in width should be reported.

Points to Watch in measuring fabric width Include

(1) The possibility of wavy selvadge due to weft tension variation or weave effects. Maximum and minumim width should be recorded and the wavy selvedge reported.

(2) Where the width "within lists" is required, the width between the innermost selvedge threads is measured.

(3) Where only samples are measured, the width of the uncoditional bulk is corrected from the measured changes in width of the samples.

Fabric Thickness

The principle of measuring fabric thickness in B.S. 2544 : 1954 : states that, 'Essentially, the determination of the thickness of a compressible material such as a textile fabric consists of the precise measurement of the distance between two plane parallel plates as the pressure foot and the other as the anvil.'

Several points require consideration in the practical application of this principle.

(i) Shape and size of the pressure foot A circular foot is usually used with a common diameter of 3/8 inch which may differ if desired. The ratio of the foot diameter to the cloth thickness should be not less than 5;1.

(ii) Shape and size of the anvil When a circular anvil is used, it should be at least 2 inch greater in diameter than the presser foot. In addition where the sample is greater than the anvil, it is convenient to surround the anvil with a suitable support for example a smooth plane board.

(iii) Applied Pressure Preferred pressure with suitable weights may be added to the presser-foot to obtain these pressures.

(iv) Velocity of Pressur Foot The presser foot should be lowered on to the sample slowly at about sec. 2/1,000 in/sec. Precision here is difficult, of course, but common sense will suggest a slow and careful movement.

(v) Time: The thickness is read from the dial of the instrument when the easily visible movement of the pointer has stopped.

(vi) Indication of Thickness A clock-type dial gauge is usually built into a thickness tester. It should be rigidly mounted in a suitable frame and after setting to zero, be capable of measuring to an accuracy of 1 per cent for cloth of 5/1,000 inch or more, and to 0.00005 inch for thinner fabric.

Fabric Weight Per Unit Area and Per unit Length

The method of obtaining weight per unit area is implicit in the titles that means, one has to weigh a known area and divide the weight by the area.

Some quadrant balances have one scale graduates in ounces per square yard. A template is used to cut the sample and the square of fabric suspended on the hook of the balance. For quick checks, this methods is useful and for particular purposes may be considered sufficiently accurate.

In weight per unit length, the minimum length measured should be 18 inches.

Conversion of Values

It is conditional provided that it is assumed that the effect of selvedge construction is negligible. Weight per unit area can be readily converted to weight per unit and vice-versa.

Let W = Weight per square yard
W = Fabric width in inches
R = Weight per running yard

Then W = 36 R/W and R= w/36

Thread Counts Per Inch in Woven Fabrics

In the woven fabric, the warp yarns are commonly referred to as "ends" and the number of warp-threads per inch width of cloth states as so many ends per-inch. The threads weft are called "picks". A fabric may threfore be described in terms of "ends and picks". For exmple- A cotton poplin may be woven with 144 ends per inch and 76 picks per inch. If the warp and weft are doubled yarn, they are called as 2/100 s. A short description of this poplin would be 144 × 76, 2/100 × 2/100 and to a person conversant with poplins, a good quality would be indicated. The determination of number of threads per inch may be made in several ways, 5 of which are given in Textile Institute Tentative Specification No-28,1995. They are:

(1) One inch counting glass. It is a simple-microscope.

(2) Traversing thread counter. It is a travelling microscope filled with a pointer to aid counting.

(3) Fabric dissection– In this method, a known width is unravelled and the threads counted. It is a useful method where the threads are difficult to distinguish, as in felted threads or where the structrue is complex. For example plied fabrics.

(4) Parallel line gratings— It is a rapid optical method.

(5) Taper—line gratings. This is a development of method No.4

The 1 inch counting glass is not recommended when the number of threads per inch is less than 25. In such cases a 3 inch sample could be unravelled and the threads counted. A ground glass-plate illuminates from below forms a useful surface on which the cloth can be laid while counting threads with a counting glass. It is sometimes convenient to count the number of repeat of the pattern instead of the individual threads. The threads per inch are then obtained by multiplying the number of threads per repeat by the number of repeats, adding on any fraction of a repeat by counting the remaining threads individually. This method is useful when the number of threads per inch is high. The region near the selvedges should be avoided because the specing of the threads in the region is often a little different than in the body of the cloth.

In the specification, it will be noted that the specimens should be conditioned for at least 24 hours before testing. This must, of course, be stated in a standard, but in practices the information is often granted quickly and this refinement is dispensed with.

Crimp of Yarn in Fabric

Definition of crimp is "to press into small folds, or corrugate". When warp and weft yarns interface in fabric they follow a wavy or corrugated path. Crimp percentage is a measure of this waviness in yarns. Peirce states that "crimp, geometrically considered, is the percentage excess of length of yarn axis over the cloth length". Percentage crimp is defined as the mean difference between the straightened thread length and the distance between the ends of the thread while in the cloth, expressed as percentage. Crimp amplitude is a term used in the literatue of cloth geometry. It refers to the eatent to which thread are deflected from the central plans of the cloth. In figure below the geometrical interpretation of crimp and crimp amplitude is illustrated using the assumption that the yarns are circular in cross-section and remain so when in the cloth.

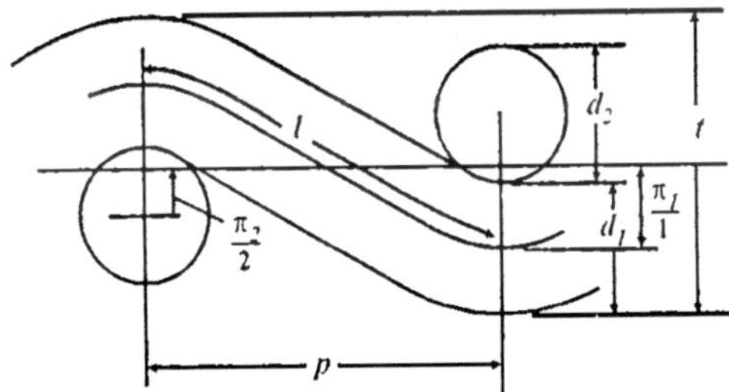

Fig. 1 : Warp crimp amplitude is the extent to which threads are deflected from the central plane of the cloth.

Air-Permeability

Due to the manner in which the yarns and fabrics are constructed, a large proportion of the total volume occupied by a fabric is, in fact air-space. The distribution of these air-space influences a number of improtant fabric properties such as warmth and protection against wind and rain in clothing and efficiency of filtration in industrial cloths. Some important vocabularies under this aspect are:

1. Air Permeability of a fabric is the volume of air measured in cubic centemeters passed per second through $1cm^2$ of the fabric at a pressure of 1cm of water.

2. Air-resistance of fabric is the time in seconds for $1cm^3$ of air to pass through 1 cm^2 of fabric under a pressure head of 1 cm of water.

3. Air-porosity is a farbic property and has the same meaning like that of the air-permeability. (2) The porosity of the fabric is the ratio-of airspace to the total volume of the fabric expressed as percentage. (skinkle).

Thermal Property of Fabrics

Measurement of the "warmth" of fabric perhaps depends on the function of the air-space and it's distribution in the structure and also on the thremal conductivity of the fibres used. Marsh trested many type of fabric and plotted their. "Thermal Insulation Value" (T.I.V.) against their thickness and obtained a fairly linear relationship thus showing that the air-in the fabric plays most important part. Cassie observes that the air is not "entrapped" as many people tend to think, but cling to the fibre-surface. Fact that probably explains the excellent properties of wool fabrics as wool-yarns, are open yarns and expose maximum fibre-surface.

Various test methods and several units have been put forward which may express the thermal properties of a fabric numerically. Some of these values are determined in nominally still air, whereas in practice, the warmth value of a fabric is appreciated most in windy conditions.

The T.I.V. mentioned above is the percentage saving in heat loss from a surface due to covering it with the fabric.

$$T.I.V = \frac{100(H_0 - H_c)}{H_0}$$

Where Ho = the heat lost per second from the uncovered surface, and

H_C = The heat lost per second from covered surface.

Fabric Stiffness, Handle and Drape

A person selecting a fabric for a given purpose usually knows what characteristics of the material are required for optimum performance in use. When a fabric is chosen for a dress material, it's technical merits if almost taken for granted and other fabric properties such as appearance, lustre, smoothness or roughness, stiffness or limpness and good or poor draping qualities are examined.

Fabric handle, as it's name implies is concerned with the feel of the material and so depends on the sense of touch. When the handle of the fabric is judged, the sensation of stiffness on limpness, hardness or softness, and roughness or smoothness are all made-use of Peirce. Drape has a rather different meaning and very broadly is the ability of a fabric to assume a graceful appearance in use. Not all fabrics. Of course, are expected to drape gracefully and indeed would be considered unsuitable if they did. One thinks of ballet skirts and stiff-petticoats.

Peirce says that the fabric stiffness is the key factor in the sudy of handle and drape. He describes. (Flexometer which is an instrument designed to measure stiffness. British Cotton Industry Research Association have produced same type of instrument for this purpose. They are 'shirley's. Combined Stiffness and Creasing Tester, a more recent version is the Shirley's Stiffness Tester. Although the use of the two instruments may differ slightly, they are both based on the same principle.

The "Shirley's Stiffnenss Tester

A rectangular strip of fabric. 6 inch × 1 inch is mounted on a horizontal platform in such a way that it overhangs, like a cantilever and bends down-ward. From the length l and the angle θ, a number of values are determined.

Bending length-C. This is the length of fabric that will bend under it's own weight to a definite extent.

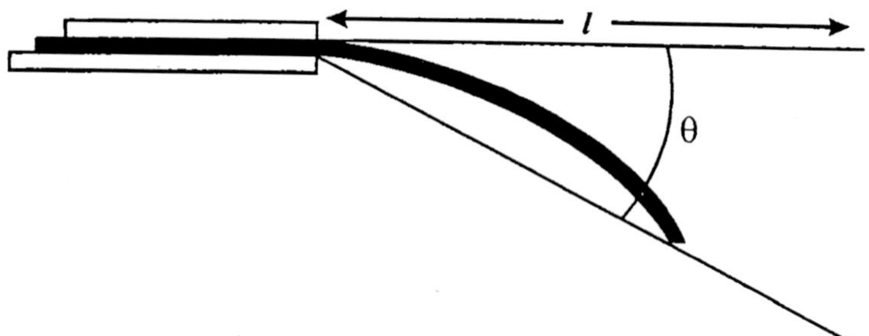

Fig. 2 : Fabric stiffness Testing, Cantilever principle

The calculation is as follows :

$$c = lf_1, (\theta)$$

Where $$-f_1, (\theta) = \left(\frac{\cos\frac{1}{2}\theta}{8 \tan \theta}\right)^{\frac{1}{3}}$$

The labour of calculating the function of θ is avoided by consulting a prepared table.

Flexural rigidity, G: This is a measure of stiffness associated with handle. A recent paper by Abbots/suggests that flexural regidity as determined by this test shows a close relationship between this value and the personal judgement of stiffness.

That calculation is as follows:

$$G = 3.39\ W_1\ C^3 \text{ mg/cm}$$

or $w_2 c^3 \times 10^3$ mg/cm

where w_1 = cloth weight in ounces per square yard and

w_2 = cloth weight in grams per square centimeter

Bending modulus q: This value is independent of the dimension of the strip-tesed and may be regarded as the intrinsic-stiffness. Peirce points out that this value may be used to compare the stiffness of the material in fabric of different thicknesses. For it's ccalculation, the thickness of the fabric must be measured at a pressure of 1lb/inch. The bending modulus is given by:

$$q = \frac{732\ G}{g_1^3}\ \text{kg}/\text{cm}^2\ or\ \frac{12G \times 10^{-6}}{g_2^3}\ kg/cm^2$$

Where g_1 = cloth thickness in thousandths of an inch,

and g_2 = cloth thickness in centimetres.

Drape-meter

The Fabric Research Laboratories of U.S.A. have developed a method of measuring drape in which the warp and seft way characteristics interact and produce the type of graceful folding seen in tailor shop window when suiting is draped over circular supports.

A circular speciment about 10 inch diameter is supported on a circular dise about 5 inch in diameter and the unsupported area drapes over-gramophone record, no draping would occur and the area of projection from the periphery would equal the area of the record. With fabrics, the material will assume some folded configuration and the shape of the projected area will not be circular. But something like as shown below:

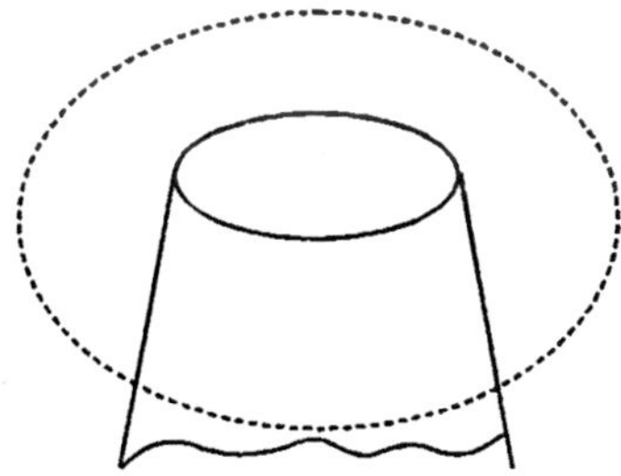

Fig. 3(a) : The circular specimen is the 'draped' part over the circular support.

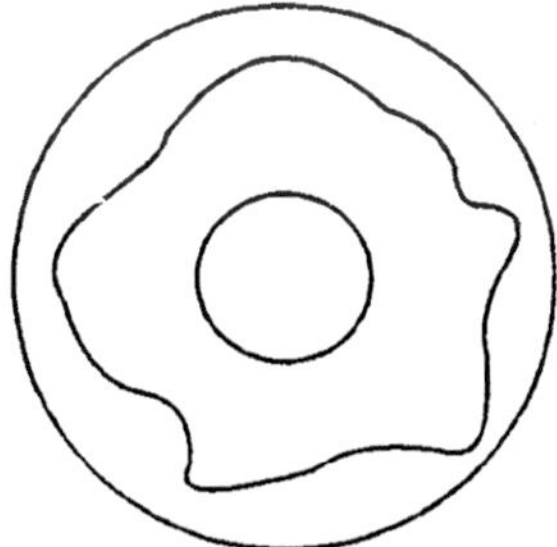

Fig. (b) : The Projected outline of the specimen.

A value known as the drape-co-efficient F, is determined by considering areas-

Let A_D = the area of the specimen
A_d = The area of the supporting disk
and A_S = The acutal projected area of the sepcimen

The drape co-efficient, F, is the ratio of the projected area of the draped specimen to its undraped area after deduction of the area of the supporting disk.

$$\text{Thus} = F = \left(\frac{As - Ad}{AD - Ad}\right)$$

Other values are obtained by further analysis of the shape of the projected outline.

Crease-Resistance and Crease-Recovery

Resistance to creasing: Cellulogic materials are notoriously susceptible to creasing and the removal of this defect may perhaps be regarded as one of the greatest achievement in the history of textile finishing.

It is important to note that many real advantages of a correctly applied crease-resistant may be accompained by a few disadvantages, especially if the finish is misapplied. Lower abrasion resistance and tearing strength have been reported and even unpleasant odours under certain conditions. Such short comings are no doubt, being thoroughly investigated by the textile chemists and will in time be either eliminated completely or reduced.

Measurement of Crease Recovery

The Total Test : This test method is a good example to demonstrate simplicity confined with effectiveness in practical application. Test specimens are cut from the fabric in both warp and weft direction, 4cm long by 1 cm wide. The specimen is folded over and creased by placing it under a strip of spring steel with a 500 g weight to supply the pressure for creasing. After-5 minutes the specimen is removed and suspended over a wire. Altogether 3 minutes are allowed for recovery after-which the distance between the ends of the inverted 'V' is measured. To avoid touching the specimen, this distance may be read from a scale engraved on a mirror below the wire.

If the fabric exhibited performance recovery, the distance between the ends would be 40 mm but, as Marsh points out, such a fabric would possess no draping qualities. Experience has shown that in order to retain good draping qualities while still having good crease recovery, the crease recovery of good-quality wool or worsted about 33 to 35 mm should not be exceeded.

The test must, of-course be on conditioned fabric and be made in a standard testing atmosphere.

Serviceability, wear and abrasion resistance The prediction of the performance of a fabric in actual use from the results of tests on laboratory instruments has up to the present cluded the textile technologist. Instead of attempting to discuss the whole problem, it is proposed to define some of the terms employed with the abone mentioned head lines.

Serviceability: An article which is serviceable is capable of performing useful service. It's serviceability ceases when it can no longer do so. The time element is included in serviceability. Serviceability-testing is done to assess suitability for the purpose.

Wear: This is the net result of a number of agencies which reduce the serviceability of an article. Some of the more important of these are bending and stretching, tearing, abrasion, laundering and cleaning.

Abrasion: It is just one aspect of wear and is rubbing away of the component fibres and yarns of fabric. Abrasion may be classified as follows:

(a) *Plane or flat abrasion:* A flat area of material is abraded.
(b) *Edge abrasion:* For example, the kind of abrasion which occurs at collars and folds.
(c) *Flex abrasion:* In this case rubbing is accompanied by flexing and bending.

The Testing of Abrasion Resistance

A number of important points require consideration before abrasion resistance tests are carried out. The choice of method may be governed by the type of apparatus available, the precision demanded and so forth. Some of the important points are outlined below:

(i) *Conditions of specimen*: Unless directed otherwise, the fabric will be conditioned and tested in a standard testing atmosphere.

(ii) *Choice of testing instrument*: The multiplicity of types has been mentioned earlier and the instrument chosen may depend upon the character of the testing desired e.g., flat abrasion, flexing abrasion etc.

(iii) *Choice of abrasive motion:* The rubbing movement may be reciprocating rotary, or multi-directional. On some instruments, the movement can be varied.

(iv) *Direction of abrasion:* When the abrasive motion is undirectional, the abrasion resistance in specific direction can be measured. In many cases difference will be observed between warp way and weft way abrasion resistance. If desired the direction of abrasion can be at right angles to the warp and weft direction.

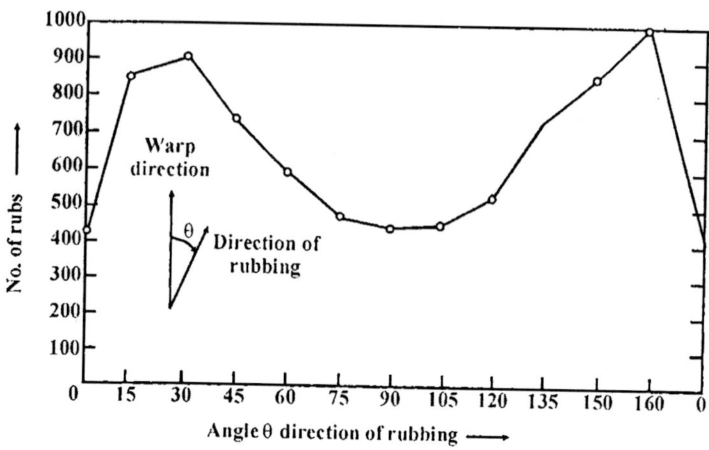

Fig. 4(a) : Effect of direction on rubbing on ball toughness test results for a viscose filament rayon satin fabric.

(v) *Choice of abradant:* The severity of the abrasion will vary with the nature of the abradant. Where possible the abrasive qualities of the material used should remain constant during the test and be capable of being reproduced for sucessive tests. Steel and silicon carbide for example will give reasonably constant abrasive qualities.

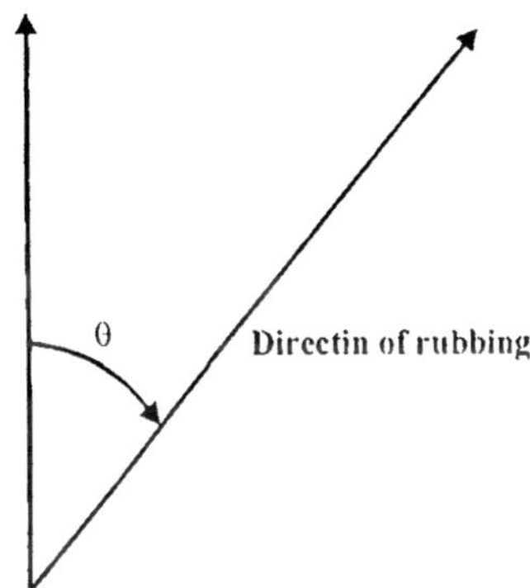

Fig. 4(b) : Angle 0 direction of rubbing.

(vi) *Backing the specimen:* The hardness of the backing of the specimen many affect the results. In some tests a hard backing is used, in others, a felt or foam rubber. In one instrument, the sample is mounted over an inflated rubber diaphragm.

(vii) *The cleanliness of the specimen and insturment:* The region to be abraded should be handled as little as possible and be free from foreign matter such as the wax from crayons or graphite from pencils. These substances tend to act as lubricants. For the same reason , the accessories of the instrument which may rub the fabric, e.g., the blade of a flex abrasion tester, must be spotlessly clean and free from grease.

(viii) *Tension on the specimen:* Standardised methods of mounting the specimen should be used to avoid errors due to variation in the tension used.

(ix) *The pressure between abradant and specimen:* The severity of the abrasion will obviously be affected by the pressure applied. Here again suitable standards must be set-up. High pressures will reduce the time taken to reach the end point of a test but the acceleration of the destruction of the fabric may lead to false conclusion.

(x) *The end point of the test:* The end point may be the completion of a given number of abrasion cycles, the appearance of a hole or broken thread, the rupture of the specimen.

Assessment of Abrasion Damage

Several methods of judging the amount of damage are used, which are as follows:

(1) Appearance against an unbraded specimen.
(2) The number of cycles, required to produce a hole, broken thread of broken strip.
(3) Loss in weight often plotted against the number of cycles.
(4) Change in thickness i.e., loss of pile height. In some cases the napping or raising effect of abrasion may cause an increase in thickness particularly in the early stages of a test.
(5) Loss in strength, e.g., tensile, bursting or tearing strength. The loss may be expressed in percentage of unabraded strength. Some laboratories may dertermine residual strength after a given number of cycles.
(6) Change in other properties e.g., air permeablity lustre.
(7) Microscopic examination of damage to yarn and fibres.

The B.F.T. Abrasion Testing Machine

The aim of the designer was to produce an abrasion testing instrument incorporating a series of mechanism to abrade textile material, which would provide numerical results that is co-related with serviceability. Examination of many rayon staple garments which had been treated with resin finishes showed that erosive wear at cuffs, collars, raised seams etc. was in the main, the type of damage which caused the garment to be classed as unserviceable or poor wearing. Flat abrasion resistance and loss of tensile strength did not appear to be major factors.

In the B.F.T. machine, special attention has been paid to the follwoing points:

(*i*) The end-point of the test is determined by the machine. It stops atomatically when the end point is reached.

(ii) The abradant used in the various accessories is made of special steel whose abrasive characteristics remain constant and are capable of being reproduced to an engineering specification.

(iii) The machine is steadily built and capable of being run smoothly at high speeds for a long time.

(iv) The results of test on the machine have been examined against a large volume of data on the serviceability of fabrics in field trials and coustomer complaints. This result can be analysed and expressed numerically and fabrics may be ranked in order of merit.

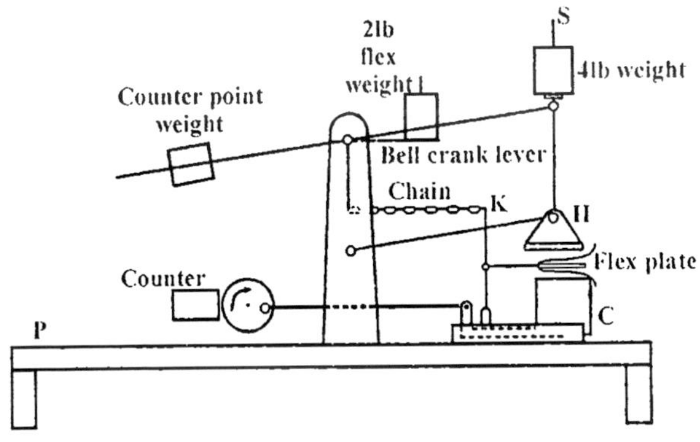

Fig. 5: The B.F.T.abrasion testing machine for flex abrasion testing.

Features of the B.F.T.Machine

The machanical parts are mounted on a rigid platfrom. P-is a solid aluminium carriage. C-is reciprocated with a one and half inch stroke at 700 rev/mins. by a crank and connecting rod, a-six-digit counter records the number of cycles. The head is mounted over the top of the reciprocating carriage and moves vertically due to the Parallal link motion. Counterpoise weights are employed so that with no weights on the spigot S, the movement is just in balance. Hence known vertical loading can be obtained by adding the required weights to the spigot. The control unit and the electric motor are mounted below the platform.

Flex Abrasion Tester

The flex is a stainless-steel plate about 0.037 on thick with one edge tapered off and sounded to 0.017 in diameter. The plate is carried in a sterrup an given a reciprocating motion. Tension is applied to the test strip by the width on a bell-crank lever acting through a chain link to the stirrup. Point *K* in the figure is a dead point in the system and here the machine can run at high speeds with flex— tension weight almost stationary, those avoiding inertia effects.

The test strip is 5 inch long by 1 inch wide. It is secured at one end over a row of stenter-pins on the carriage C and led round the radiused edge of the flex-plate the free-end is then secured over-stenter pins on the head H. A load of 4 lb is added to the spigot 8 and 2 lb weight to the bell crank lever, a 2:1 velocity ratio causes the tension, in the specimen to be 4 lb. The head H is shown lifted in the figure for sake of clarity but during the test, it rests on the top of the test specimen and flex-plate. When the machine is switched on, the reciproacation of the carriage caused the fabric to be repeatedly pulled back and forth round the edge of the flex plate and continuous flexing is achieved. Eventually the fabric breaks when a switch is operated, automatically stopping the machine. The number of cycles is read from the counter and recorded. The mean of five tests is taken-warp-way and weft way, or in both direction if necessary.

Since quite a high numbers of cycles are obtained, the mean values are reduced to more manageable proportion by using a "normalising factor", of 10^{-3} dividing the mean value by 1,000. Hence,

$$\text{Flex result P,} = \frac{\text{Mean value of fine tests}}{1000}$$

The 'Pilling' of Fabrics

Pilling is a fabric surface fault characterised by little 'pills' of entangled fibre clinging to the cloth surface and giving the garment, an unsightly appeaarance. The pills are formed during wear and washing by the entanglement of loose-fibres which protrude form the fabric surface. Under the influence of the rubbing action, these loose fibres develop into small spherical bundles anchored to the fabric by few unknown fibres. Pilling has long been recognised as a fault, especially in fabrics such as woollen, knitted goods made from soft twisted yarns, but the introduction of the newer man-made fibres appears to have aggravated it's seriousness.

As pills form due to the migration of fibres from the constituent yarns in the fabric, it follows that the reduction or prevention of pilling may be affected by reducing this migratory tendency. The methods used include the use of higher twist factors, for the yarn, the bursting and cropping of the fabric surface, and special chemical treatments are conclusion reached was that pilling is reduced by increasing the nylon filament denier. It should be remembered that whatever antipilling technique is used, the fabric must not lose it's disirable handle or other qualities.

Pilling Tests

The pilling of a fabric is a fault on the fabric surface characterized by little 'pills' of lenfangled febre clinging to the cloth surface and giving the garment an unsightly appearance. Pills are formed during wear and washing by the entanglement of loose fibres which probcude from the fabric surface. These loose fibres, under the influence of the rubbing action, develop in to small spherical bundles, anchared to the fabric by a few broken fibres.

After rubbing under controlled condition the pilling of the sample may be assessed numerically by counting the number of pills formed. Alternately, the appearance of the test specimen may be compared with standard samples and given some form of rating.

Numerical Assessment of Pilling

The instrument is the Martindale Abrasion Tester. The normal sample holders are repalced with light-weight square holder which are keyed so that they may have vertical movement but cannot trun on their axes. The samples are given a multidirectional movement and are rubbed against a standared-fabric. After a certain number of rubs, the samples are examined and the number of pills counted. This may be repeated in stage of 500 cycles up to 3,000 or 5,000, and the rate of development of pills noted.

Water and Fabric Relationships

The merit of a fabric intended for rain-wear, waggon covers, or tents is judged, amongst other properties, by it's ability to keep water out. Conversely when intended for hosepipes or canvas buckets, to keep water in. In another direction, some fabrics must exhibit the ability to absorb water rapidly low :

1. *Proof (water)(verb):* To treat textile materials e.g., with fats, waxes or rubber to prevent absorption of water. The additions may be physical films or caatings or may be physically combined. The feature of a waterproof (fabric) is the low-degree of permeability to air.
2. *Proof (Shower) (verb):* To treat tiextile materials in a manner to delay the absorption and penetration of water. The fabric retain a degree of permeability to air. By proper choices of fibre and of yarn and fabric constructions, cloth can be made which of themselves are shower-proof.
3. *Water repellent (Adjective):* This is a state characterised by the non-spreading of a globule of water on a textile material. In the 'waterproof' fabrics the fibres and the threads have been filled with the rubber or other proofing-material. Since the fabrics are almost impervious to air, such materials would be uncomfortable to wear. In practice these materials are used for outer-wear, it is usually to punch ventilation holes in the garment as for example, uner the sleeves. The term wettability is the rate of wetting of the fabric.

The Basic Concept of Wetting and Water Repellency

Suppose a drop of water is put on each of three smooth solid surface in each case it's behaviour observed. In Figure:

(a) it may assume an almost spherical shape, in

(b) it may assume some incermediate form between these two extremes.

(c) it may become nearly flat or

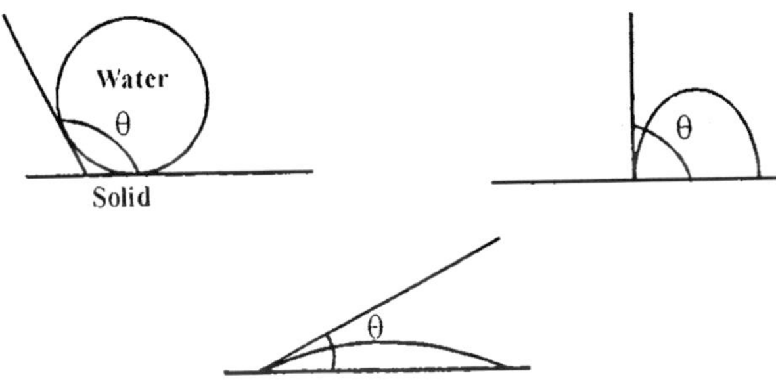

Fig. 6 : Contact angles

Contact Angles-Diagram

In the above figures, the angle θ is known as the "contact angle" and may be defined as the angle between the solid surface and the tangent to the water-surface as i' approaches the solid, the angle being measured in the water. The water properties are the differences in the behaviour of the drops of water on the three surfaces and must be due to the differences in the properties of water or surface combinations. A solid surface on which a drop of water form a shpere i.e., shows a high contact angle, will tend to shed the water, the drop will 'pearl' and roll off leaving the surface dry. Where a low contact angle occurs, the drop-spreads over the surface and "wets" it.

To illustrate these effects, two microscope slides are taken, one thoroughly cleaned and the other given a thin coating of paraffin wax. A small globule of water is placed on each. On the waxed surface, the water will become nearly spherical and when the slide is tilted the water will roll away and leave the wax surface dry. The drop on the clean slide will spread out and wet the slide.

This is due to the mechanics of surface energies and

surface tensions at interfaces, (an-interface being the surface between two media, e.g., solid-liquid, liquid-air, and solid-air etc). The mechanics of these interfacial tensions and energies are explained by Baxter and Cassie who extend the theories to the problem of wetting rough and porous textile surfaces. In this experiments it was observed that with a water repellent material, a high receding contact angle is exhibited. Contact angle may be "advancing" or receding. The Advancing contact angle is the one in which the liquid advances to a dry surface. The receding contact angle is the angle where the liquid recedes from a wet surface. They give an example of high advancing contact angle coupled with a low receding contact angle, the angle shown by a drop of rain rolling down a dirty window pane.

Wetting of Fabric in Rain

A drop of water falling in to a fabric will penetrate most deeply in to the pores between the yarns. The ability of the water to withdraw form the pores depends on the yarn-water receding contact angle being high.

With a reasonably proofed fabric, the initial yarn receding contact angle will be high and this ensures that the water will contract out of the pore in to the drop and pearl off. Succeeding drop falling on the pore will then force the water further in to it and the water will withdraw to a less extent because the yarn-water receding contact angle has decreased. This process of penetration of fabric by rain will continue until a kinetic energy of the falling drops is not sufficient to force the water column deeper into the pore. When this condition is attained, the absorption of the fabric will have reached a limiting value and further exposure to rain will result in no further absorption. If the thickness of the fabric is greater than the maximum depth, penetration of the water through the fabric will take-palce. Thus the maximum absorption of a fabric and the penetration are determined by the fibre-water advancing contact angle and the physical construction of the fabric, while the rate of picking up of water depends on the effective receding contact angle θ and it's decay with time. It is necessary to conclude that a practical test for water repellency should be influ-enced both by the effective advancing and receding contact angles as both are equally important in the wetting of a fabric in rain.

Methods of Testing

This test was developed by Baxter and Cassie following their research on water repellency. The apparatus used throws an enlarged image of a strip of fabric, end-on, as it is slowly withdrawn from the surface of water in a small tank. Distilled water at 20⁰ C is used and the speed of withdrawal is 8 mm/min. At the start of the test, a large receding contact angle is seen but after a time, noted by a stop-watch, the angle decreases to 90 degree. The time taken to drop to 90 degrees is called "wetting time". In a series of comparative test, it was found that this method was the most sensitive to assessment of proofing efficiency on heavy-wool cloths, whereas on the cotton fabrics, all three methods used, such as wetting time, hydrosbatic head, and the Bundesmann, ranked the fabrics in roughly the same order.

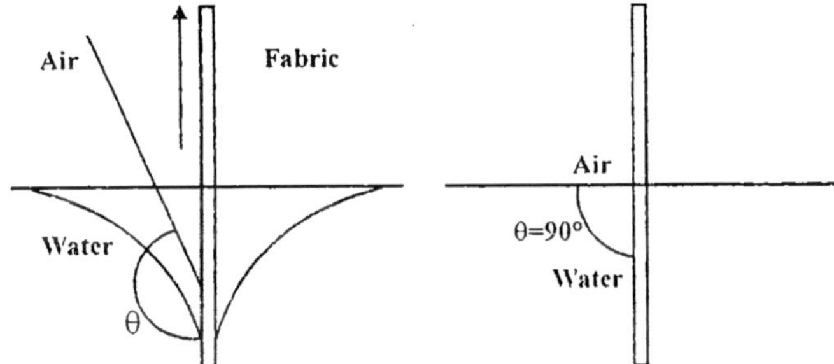

Fig. (7) : The 'Wetting-time' test

Wettability of Cotton Fabrics

The contact angle is again considered in this test developed by the British Cotton Industry Research Association. A drop of water (or sugar solution) is placed on the specimen fabric which is mounted horizontally. The time taken for the contact angle to drop to 45 degrees is noted. The reciprocal of the time taken is called the "wetting velocity" or simple "wettability.

The Spray Test

In this test, a small scale mock rain shower is produced by pouring water through a spray nozzle. The water falls on to the specimen which is mounted over a 6 inch diameter embroidery hoof and fixed at an angle of 95 degree.

To carry-out this test, 250 cm of water at 70° F are poured steadily into the funnel. After spraying has finished, the sample-holder is removed and the surplus water is removed by tapping the frame six-times against a solid object, with the face of the smaple-facing the solid object. The tapping is in two stages, three taps at one point on the frame, and then three times at a point diametrically opposite. The assessment of the fabrics water repellence is given by the "spray-rating". After removing the surplus water the fabric surface is examined visually.

The American Association of Textile Chemists and Colorist recommended the use of a chart of photograph against which the actual fabric appearance compared. The ratings are as follows :

100 = No sticking or wetting of the upper surface.

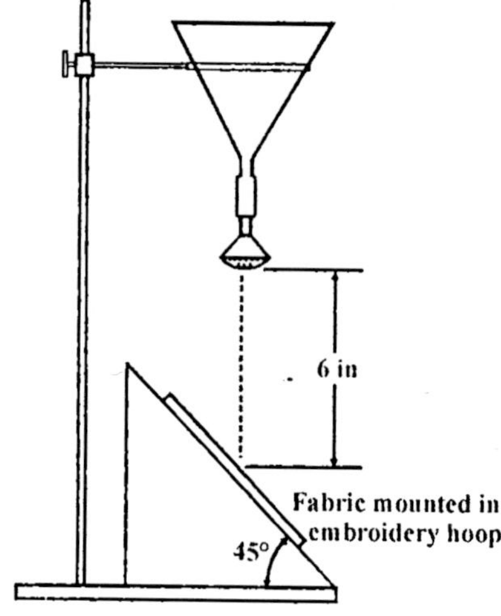

Fig. (8) : The spray-test

90- Slight random sticking or wetting of the upper surface.
80- Wetting of upper surface at spray-point.
70- Partial wetting of whole of upper surface.
50- Complete wetting of whole of upper surface.
0- Complete wetting of whole of upper and lower surface.

The Drop Test or Drop Penetration Test

It was noted earlier that in the initial stages of wetting, the drops of water pearl off the fabric, but in time pearling ceases, the fabric become wet. The drop test is a count of number of drops required to penetrate through to the underside of the fabric when all the drops fall on the same spot.

In this experiment, the fabric-specimen is clipped on to a glass-plate with a piece of filter paper sandwiched between the fabric and the glass. A frame holding the assembly at an anle of 45 degrees–directy under the drop-forming device. The later is prepared from a fine-bore-glass tube to produce a certain number of drops of a given size in a minute, with a constant head of water. To ensures that the drops fall on to the same spot, a drought shield of large diameter is used.

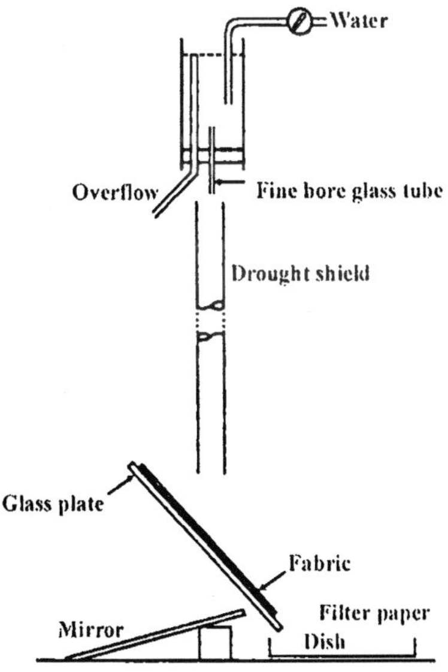

Fig. (9) : The drop penetration Test

With the specimen in portion, the water-supply is started and the drops begin to fall on the fabric. The endpoint is reached when the filter paper shows signs of water, a mirror underneath the specimen assembly facilities observation. Various methods have been tried to determine the end-point with greater precision.

The Pentration of Fabric by Water under Pressure

The pressure required to force-water through a fabric may be determined and the information used is the assessment of the fabrics ability to do a particular job.

Sinking Test

This involves a simple test of wettability of fabric. In this test, a small square specimen about 1inch × 1 inch is cut and drap it to the surface of water in a beaker. The time taken for the specimen to sink below the surface is observed. The shorter the time, the greater the wettability.

Wetting by Wicking

The ability of fabric to absorb water, especially by wicking or capillary action may be observed by timing the rate at which water climates up a narrow strip of fabric suspended vertically with it's lower end dipping in to the water. In this experiment a square of fabric 2 inch × 2 inch is fully immersed in water at 20^{0}C for 20 sec. removed, and weighed. From it's known dry weight the percentage water absorbed is calculated. Garner-states that good towelling should absorb 100% of its own weight.

Shrinkage Tests

A fabric can have great technical merit if its dimensions remains constant throughout its useful life. This dimensional change occurs when a fabric is washed. And after few washing some fabrics shrinks. Test for shrinkage involves.

(i) Preparation of specimen by conditioning in a standard testing atmosphere and marked with usually three pairs of datum lines in each direction with indelible ink or fast-dyed sewing thread.

(ii) washing the sample in relevant washing solution in a washing machine confirming to certain specification and ringing at the sample after specified time.

(iii) Drying after removing surplus water by centrifuge or by hand screening, rubber cover roller wringer, or rolling in towelling. Drying is completely by meaning of a flat-headed press or a heated-flat iron.

(iv) Conditioning and re-measuring the specimen in a standard tasting atmosphere and measuring the distance between the dalum lines.

Percentage shrinkage is calculated from the mean changes in the distance between the datum lines.

$$\text{Percentage} = S = \frac{100(L_0 - L_1)}{L_0}$$

where L_0 is the distance between the datum lines before washing and L_1 is the distance between the datum lines after washing. Carpet testing. The major properties of carpet are (1) durability (ii) retention of appearance (iii) Pile height and density. To take the lost property first, the density of the pile may be derived from a knowledge of the pile height and the carpet construction. The various tests for carpet are (a) Durability test which investigates the length of life of the carpet before backing becomes visible due to fibre tube loss.

(a) Where the sample is undergone with heavy beating and rubbing action simulating the actual action of walking in tetrapod walker carpet testing where the drum rotating with the sttetopointed tetrapod moving include on the tiffs of carpet.

(b) Appearance and texture retention test where the carpet sample is subjected to beating and rubbing action for sufficient time to bring about a substantial change in appearance. A similar test is also done on the sample in damp condition to observe the retention of appearance.

(c) Soiling test—The soil rustance of carpet is tested placing a measured amount of artificial soil in the drum of tetrapod tested and the drum is rotated for 10,000 revolution along with the sample inside. Then the carpet sample removed and examined for the degree of soiling.

(d) Resilience test—The ability of carpet to recover from the compression of walking traffic, furniture feet etc. is a desirable property. Studying the dimensional changes of a carpet can asses the residence property. Measuring the thickness of a carpet before or after tetropod action asking conventional thickness gauge can assess the resilience reiteration in relation to pile fatigue.

(e) Carpet thickness—A carpet is compressible and therefore it is necessary to specify at what pressure its thickness is to be measured. Measuring the thickness of a new carpet and then measuring it after use or test can determine the loss of thickness.

6

THE TENSILE TESTING OF TEXTILES

The strength of fibres or fibre structures is commonly regarded as the criterion of quality. Although, strength in testing is of great importance, but properties such as flexibility, resilience, moisture absorption, dye affinity etc., are necessary to be considered in the assessment of goodness or quality of a textile material. It should always be kept in mind that the relative importance of one particular property will depend mostly on the end use. It is for sure, the measurement of tensile properties of textiles is an important branch of textile testing. Prior to getting through the methos and principles of this testings, the different units and terminologies employed are very essential to be defined and understood clearly. There are :

Load — This refers to the application of pressure to a specimen in it's axial direction. It causes a tension to be developed in the specimen. The load is usually expressed in grams weight or pounds weight like gravitational units of force. it has become common practice to leave out the word 'weight' and quote loads in grams on pounds.

Breaking Load: This is the load at which the specimen break usually, expressed in grams weight or pounds weight.

Stress: Here, the merits of fine and coarse structures have to be compared, suitable units must be chosen. In engineering the term "stress" is used. This is the ratio between the froce applied and the cross-section of the specimen.

$$\text{Stress} = \frac{\text{Force - applied}}{\text{Cross - sectional area}}$$

The force is accurately expressed in dynes or pundals. Hence, the stress on a fibre may be given in terms of dynes per square centimeter (dyn/cm^2).

Mass-stress

The cross-sections of many fibres and fibre structures are irregular in shape and difficult to measure. To simplify matters, a dimension related to cross-sections is used, known as the "linear density" of the specimen. The linear density may be expressed in denier or tex-count and the "mass-stress then becomes the ratio of the force applied to the linear density (mass per unit length)

$$Mass\ stress = \frac{\text{Force applied}}{\text{Linear density}}$$

The units of the mass-stress therefore becomes grams weight per denier or grams weight per tex. Here again abreviations are used: "Mass-Stress" is usually shrotened to stress and quoted in grams per denier or grams per tex.

Tenacity or Specific Strength

The 'tenacity' of a material is the mass stress at break the units being of course grams per denier per tex. An alternative term for tenacity is "specific strength". The value used for the denier of a specimen is usually the nominal denier before testing, no account being taken of the decrease in denier as the specimen is stretched and becomes finer.

Breaking length: This is the length of the specimen which will just break under its own weight when hung vertically.

For example: a 100 denier viscose rayon yarn broke at a load of 185 g, the breaking length would be:

$$\frac{\mathbf{185 \times 9000}}{\mathbf{100 \times 1000}} = \mathbf{16.65}$$

Strain: When load is applied to a specimen a certain amount of stretching takes place. The amount will vary with the initial length of the specimen. The 'strain' is the term used to relate the stretch or elongation with the initial length :

$$\text{Strain} = \frac{\text{Elongation}}{\text{Initial length}}$$

Extension: By expressing the strain as a percentage we obtain the "extension" :

$$\text{Extension} = \frac{\text{Elongation}}{\text{Initial length}} \times 100 \text{ per cent}$$

The extension is sometimes referred to as the "strain percent".

Breaking Extension

The breaking extension is the extension of the specimen at the breaking point.

The load elongation curve : when the load on a specimen is plotted against elongation, an extremely important curve is produced. This curve describes the behaviour of the specimen from zero load and elongation up to the breaking point.

A close study of this curve can yield further important information like initial Young's Modulus, work of rupture, yield point etc.

The stress-strain curve : To enable more direct comparisons, to be made between different types of material and structures, it is more convenient to convert the load elongation curve to a stress-strain curve. To convert the curves to stress and strain curves we must draw up a table of values of stress and strain derived from measurements on the load-elongation curves. Stress-values are obtained by dividing load values by the denier and the strain values are calculated by dividing the actual elongation by 20 in the test length. The figure 1(a) below shows the load elongation

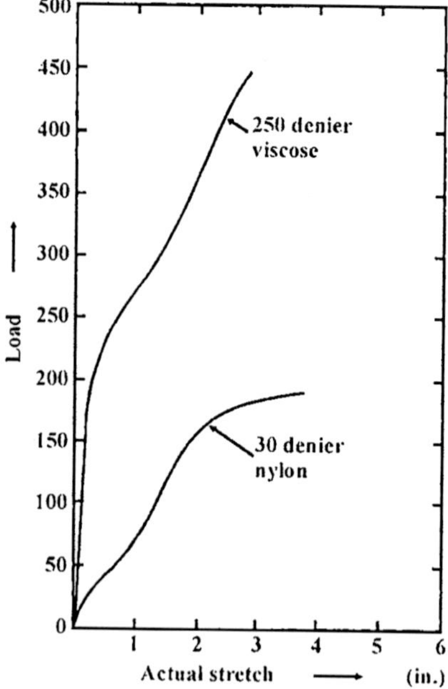

Fig. 1(a) Load-extension curve of nylon and viscose obtained on scott serigraph.

curves of a 250 denier viscose rayon yarn and 30 denier nylon yarn. If at 1 inch elongation, the load on viscose was 272g, the stress at this point would be 272/250, 1.08 g/denier. The corresponding strain was 1/20 or 0.25. Figure 1(b) shows the stress-strain curves derived from load elongation curve. The general shape of the curve remains the same but their relative position have changed. The superior strength of nylon is more clearly seen and the comparision between the two types of fibre made easier.

The Initial Young's modulus : Always the initial portion of the stress-strain curve starts at zero strain and stress which is an important part of it. It can be observed from figure 1(c) that the first portion of the curve is fairly straight. This indicates a linear relationship tested on a Scott Serigraph which operates on the inclined plane principle to give a constant rate of loading between the stress and strain. This port of the curve is sometimes referred to

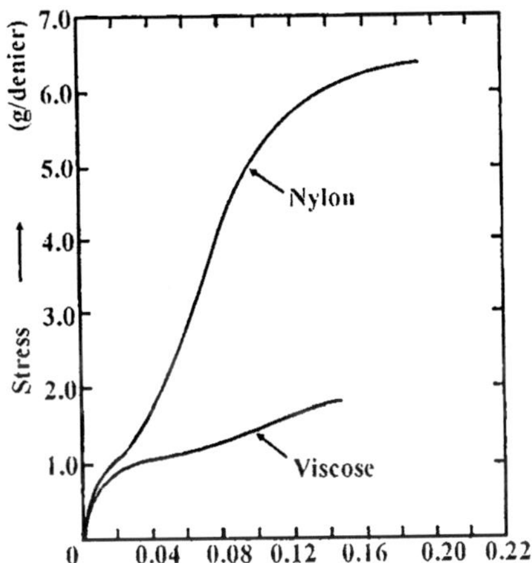

Fig. 1(b) : Stress-strain carves obtained from the load-extension curves of nylon and viscose.

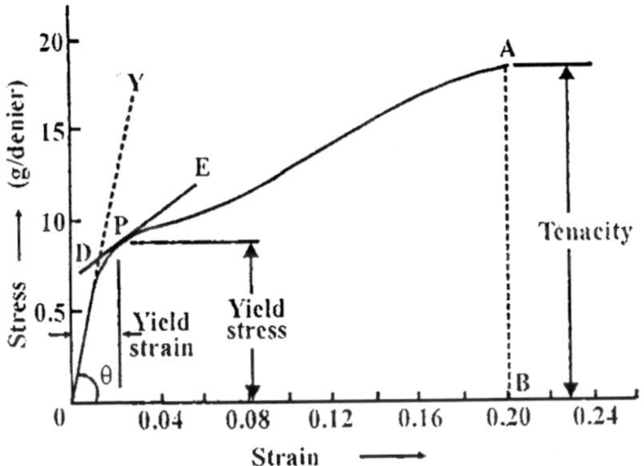

(1) DE | QA, P is the yield point.
Therefore, Yield stress = 0.88 g/denier
Yields strain = 0.02

(2) Initial Young's modulus = $\tan\theta = \dfrac{1.66}{0.02} = 83$ g / denier

(3) enacity at break = 1.85 g/denier

(4) Extension at break = 0.20 × 20 per cent.

Fig 1(c) :

as "Hookean Region". It signifies that when the load is removed, the material recovers its original length or almost nearly to it. The angle between the initial part of the curve and the horizontal axis is the ratio of stress - strain. And this ratio in engineering science is termed as the Young's Moudulus, which gives a measure of force required to produce a small extension. This initial Young's Modulus describes the initial resistance to extension of a textile material. The initial Young's Modulus will be in grams per denier if the stress is in units of grams per denier. It is because the strain is a fraction and has no units. The Initial Young's Modulus can also be expressed in Kilometers weight of the material considered.

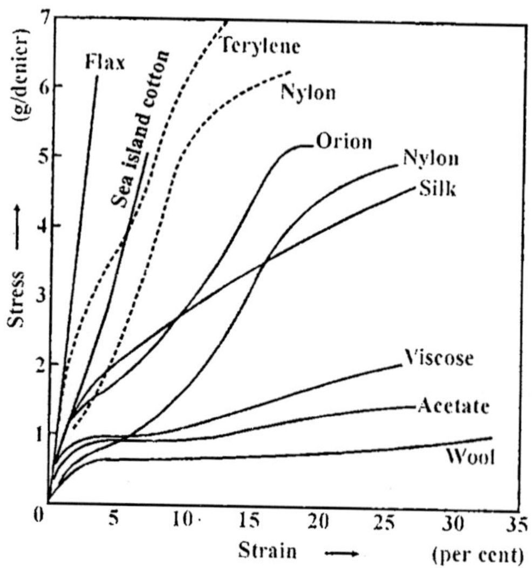

Fig. 1(d) : Stress strain curves for various materials. Note : These curves are intended for general comparisons only. The properties of man-made materials can be varied by special processing techniques and their corresponding stress-strain curves are changed. Thus the yield point can be defined in terms of the yield stress and the yield strain. In an other context the yield point is the 'limit of proportionality' where the extension ceases to be proportional to the stress and 'elastic limit.'

Yield Point : It is important to note that the straight part of the curve gives way to a bend beyond which the material no longer behaves like a spring — or elastic. Relatively small increases in stress produce large extensions. And most of the stretch is irrecoverable. This bending or yielding region which is determined geometrically is known as yield point. This is the point at which the tangent to the curve is parallel to the line joining the origin and the breaking point.

The 'Work of Rupture' : This refers to a measure of the toughness of the material. It is the energy or work required to break the specimen. The area under the load-elongation curve represents the work done in stretching the specimen to breaking point. Thus the units of the work of rupture is presented in units of work as gram-centimeters. It should be clear that the work of rupture is proportional to the cross-section of the specimen and it's length. So to make the comparison easier and perfect, the work of rupture may be expressed in gram-centimeter per denier per centimeter length:

$$\text{Thus} \quad \frac{\text{Gram centimeter}}{\text{Denier per centimeter}} \quad \text{or} \quad \frac{\text{g.cm}}{\text{den.cm}}$$

Work Factor : As per the Hooke's Law, the stress-strain curve from zero load to breaking load, would be a straight line.

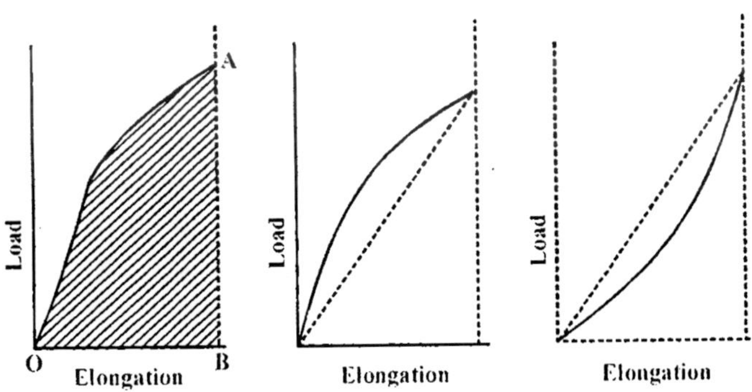

Fig. 1(e) : The work of rupture and work factor : (a) Work of rupture = area OA; (b) Work factor > $\frac{1}{2}$; (c) Work factor < $\frac{1}{2}$.

The ratio of this curve, that is

$$\frac{\text{Area under the curve}}{\text{Breaking stress} \times \text{breaking strain}}$$

Would be equal to one-half. This ratio for a particular curve is known as the work factor. When the curve is concave towards the strain axis or when it is horizontal, the work factor is greater than one half. And in the other way, if the curve is concave towards the stress-axis the work factor is less than one-half.

Elastic Recovery : The term elasticity refers to that property of a body by which it tends to recover it's original size and shape after deformation. The power of recovery from a given extension may be expressed by the Elastic Recovery value as the following—

$$\frac{\text{Elastic Extension}}{\text{Total Extension}}$$

To be more clear, if for example the original specimen length is AB and stretched length is AD, the total extension would be BD. The length may be come AC after removal of the load. Thus, the length CD is the elastic extension. So—

$$\text{Elastic recovery} = \frac{CD}{BD}$$

A B C D

AB = Original length CD = Elastic extension
BD = Total extension BC = Permanent set.

$$\frac{CD}{BD} = \text{Elastic recovery.}$$

Fig. 1 (f) Elastic Recovery.

Instantaneous and time dependant effect :

The behaviour of a specimen while study can be split in to two :

(i) The instantaneous effects.

(ii) The time-dependant effects.

When a load is applied on a specimen constantly for a period of time, the specimen initially extends repidly giving the instaneous effect and then less rapidly which yields to the time dependant effect. After removing the load from the specimen, the recovery of extension is rapid at fixed and the slowly may be with a small amount of residual extension known as 'permanent set'.

The instantaneous extension is composed of two quantities of extension,

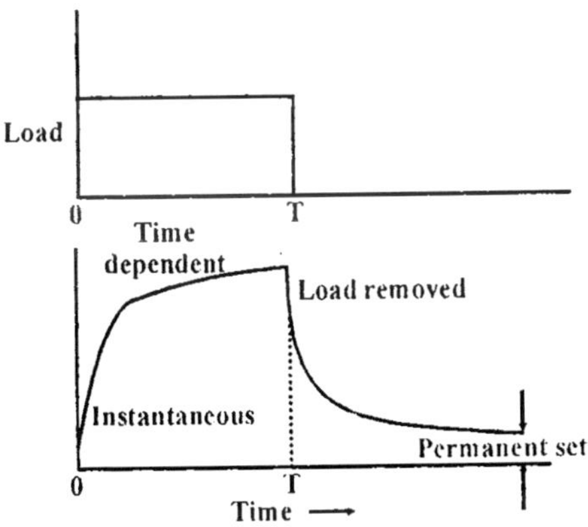

Fig. 1(g) : Instantaneous and time dependent effects.

(i) The elastic extension which can be recovered and.

(ii) The plastic or permanent extension which cannot be recovered. The time dependant behaviour also can be considered as (i) Primary creep which recovers back and the (ii) Secondary creep which does not.

Loading the specimen beyond the yield point—This process is advantageous in the processing of filament where the tension adjusted to valves which are below the yield point. Over stretched yarn becomes lustrous and results with a fault known as 'shiners' in the fabric. This test involves loading a 300 denier viscose yarn first with 200 gm. and then unloading to zero again. In the figure below, the stress is 0.67 g/denier at point X which is below stress of approximately 0.85 gm/denier. This is the reason for which most of the total extension OA is recovered at zero load and with nearly 1.0 elastic recovery. Again the same yarn is loaded up with 360 gm with 1.20 g/denier stress. Now the point Y is found to be well-post the yield point P. This time the total extension is OD. After the removal of the load the yarn recovers some of the extension until at zero load where the extension was OC. When the yarn in the test was taken to it's breaking point the final stress-strain curve CE has a different shape than OY. This shows that as per the mechanical conditioning by stretching previous history of a specimen can alter it's tensile behaviour.

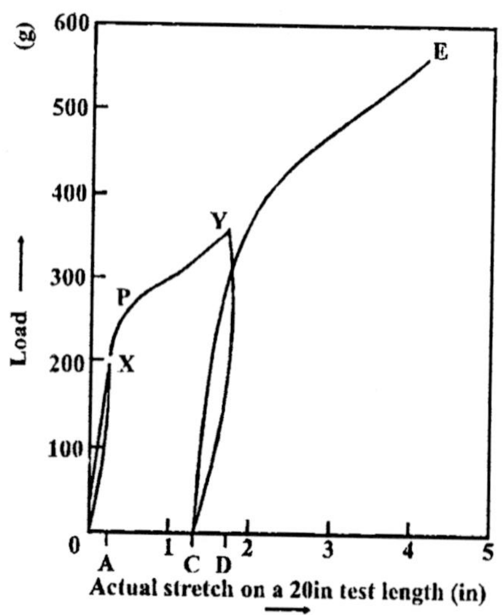

Fig. 1(h) : Load-extension curve for 300 denier filament viscose yarn showing the effect on elastic recovery when the specimen is loaded beyond the yield point.

Elastic recovery and dimensional stabury in clothing, dimensional stability is the ability of a garment to keep it's shape. This aspect also involves elastic recovery. A suit made from a material with poor elastic recovery property would exhibit baggy knees in the trousers and baggy albows in the jacket.

Factors affecting the tensile properties of textiles

(i) Length of the test specimen : There is a desirable relationship between the strength, test length and irregularity while testing the tensile property. As per the processing point of view, a yarn with a slightly lower mean strength but of greater uniformity is preferable to a yarn of higher mean strength but irregular in appearance and finish. The incidence of weak places is one cause of excessive end breakages in winding the weaving etc. The more irregular the material is, the greater will be this weak effect.

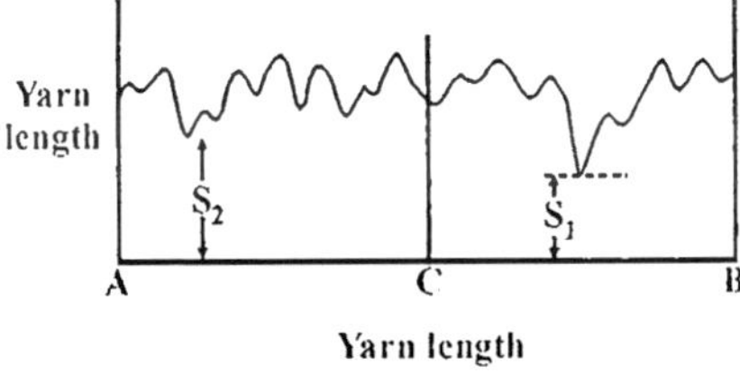

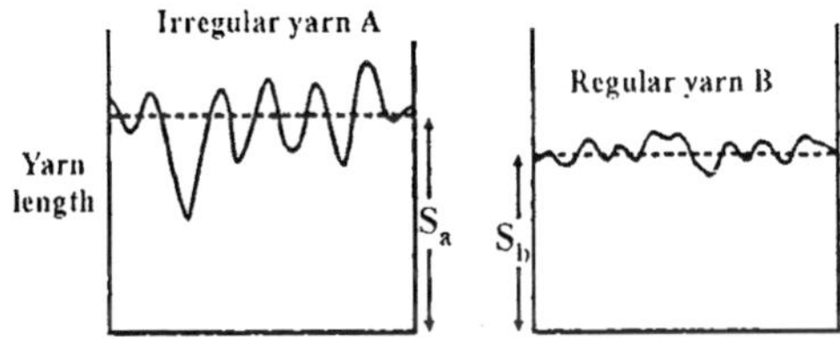

Fig. 1(i) The 'weak link' effect : (a) The strength or 'breaking load'; (b) Yarn A would give a higher mean strength, S_a, than yarn B_2, S_b, if tested over short lengths, but would give a higher end-breakage rate in processing due to incidence of very weak places.

In this, an irregular yarn a and a regular yarn B were compared with respect to their single thread strength properties. The regular yarn B would produce the higher strength, if the test length chosen were long. But with shorter test lengths the mean strength of yarn A would approach the value SA, which is a higher value than Sb. This, it would be possible to reverse the ranking of the two yarns by adjusting the test lengths.

(ii) Rate of loading and time to break the specimen : A rapid test produces a higher breaking load than a slow test. There is an established empirical relationship between the strength values obtained and the time taken to break the specimen. As this relationship appears to apply to yarn and fabrics as well as to fibres, it is clear to show how the yarn and fabric properties depend mainly on fibre properties.

(iii) Capacity of the machine : The capacity of the machine should be chosen so that the time required to break the specimen is close to the recommended time. If a weak specimen is tested on a high capacity machine, the time to break it will be short and therefore an optimistic strength result will be produced.

(iv) The effects of humidity temperature : The amount of moisture in the specimen influences the mechanical behavour of textile fibres and fibre structures. Always the various types of fibre differ in their moisture relationship. So the degrees to which the fibre properties are modified will vary automatically.

(v) Previous history of the specimen : It is always desirable to know the previous history of the specimen to avoid the results which leads to errors. It is always clear that the behaviour of a specimen after being strained beyond the yield point differs from it's behaviour when not strained. Besides, treating with chemical may also affect the tensile properties of the specimen. Damage to the fibre by chemical may result with seriously tendering the fibre.

(vi) The form of the test specimen : The complexity of structure in many test specimen may range from complicated structure of single and plied yarn to woven and knitted fabrics and other forms of manufactures. Change in the yarn strength, elasticity liveliness, luster and other yarn characteristics are caused due to the changes in the twist factors.

Yarn Strength Testing

There are a wide variety of instruments available for testing the strength of yarns. They may be simple, some are complicated, others have recording devices and some of them have autosystems. It is often difficult to choose the best type. But for comparing the various methods of yarn testing, a purely technical angle is very useful.

The lea test—It is conducted for the older sections of wool, cotton and flax on a pendulum lever tester. It is the standard method of measuring the 'strength' of the yarn. The fractional forces rendered to the yarn in the lea test reduce the sensitivity of the test to detect weak places in the yarn. A weak yarn will produce a low lea strength, but presence of an abnormally weak place in one of the treads may not be detected.

The count strength product—From to strength point of view, the product of the count and lea strength is a useful measure of the merit of the yarn. It is essential while assessing the characteristic of yarns. The count strength product (CSP) enables comparison to be made between the yarns of similar but not necessarily identical count (CSP is obtained by multiplying the count of yarn with the lea-strength) CSP varies as the twist factor is Changed. The spinning quality of a cotton or the spinning efficiency of a particular system may be judged from the concept of CSP. The finer the count of a yarn, the smaller the CSP or the CSP falls comparatively. At certain count, the CSP will be equal to the standard. Then this count is termed the 'Highest Standard council' (HSC). Thus, the HSC for a sample of cotton is a measure of its spinning quality.

Skein—breaking tenacily—Skein is a specimen size for yarn count. For testing the tenacity of yarn, the specimen size for yarn count is a skein of 50 m long, that is 50 wraps on a one meter (1m) wrap reel. this skein will be tested on a strength testing machine and the breaking load measured in Kilogrames. In palace of count × strength product (CSP), a measure of yarn strength would be calculated as follows:

Skein breaking ten acity (grammes per tex) =

$$\frac{\text{Average skein - breaking load in kilogrammes} \times 1000}{2 \times \text{wraps in skein} \times \text{Average liner density by unstrained yarn in tax}}$$

A direct comparison of this figure with single thread tenacity from single thread tests would be possible.

The ballistic test : This test records the amount of energy or work required to break a hank of yarn and in effect three characteristics of the yarn are combined in one figure that is, the breaking load, breaking extension and work factor. The breaking load and the breaking extension cannot be separated. Besides, a strong yarn which cannot be extended may not be distinguished from a yarn which is weaker but more extensible. The ballistic test is sensitive enough to detect changes in yarn quanlity and machinery performance as well. It also can detect increase in yarn irregularity more sensitively than the lea test.

The single—thread test—In order to eliminate the disadvantages of lea test, single-thread test is preferred. The test specimen for single spun yarn is 50, with a minimum number of 20 is used in the standard test procedure. A measure of irregularity of the material is obtained, by calculating the co-efficient of variation of the strength and the extension. There is no simple relation between the single thread strength and the lea-strength. The strength per thread for the lea test may be calculated by dividing the lea strength by the number of threads between the hooks. For example divide the hank is 120 yd. Hand would in 80 revolutions of the wrap real divide by 160. This value of strength per thread can be expressed as a percentage of the average single thread strength.

Standard-single-thread strength value—To compare ane's own product with those of the other yarn producer, the manufacturers of the 'Uster' tester have prepared a set of standards based on the results of tests on many yarn from different countries. These standards based on the results of tests on many yarn from different countries. These standards are for carded and combed cotton yarn spun rom various raw cotton varieties and at different twist factors. The breaking strength is expressed in terms of "breaking length". Besides, the count is expressed in tex but the twist factor in English form. For example, a carded cotton yarn, spun from Texas cotton, 11/16 in. Stable, to a count of 24s (English) with a twist factor of 4.00, should give breaking length of 14.25 km.

Fabric Strength Testing—Strength of fabric is the essence of textile without which the value is nothing. It can be ascertained by certain tests. There are numourous reasons stone of the common reasons for carrying out tests on fabrics are—

(i) To check that the fabric conferms to specification.
(ii) To note the effects of changes in structural details.
(iii) To note the effects of physical and chemical treatment exposure to weather, laundering etc.
(iv) To obtain some indication of probable performance in use.
(v) To investigate causes of failure and customer complaints.
(vi) To help in design of a fabric for a specific purpose.
(vii) To study the inleraction of fibre, yarn and fabric properties.

Some of the main tests are :

Strip Test : For conducting strip test, the specimen used in the strip test is usually a 2 in wide piece of fabric pepared by initially cutting the material to a width of about 11/2 in and removing threads from both edges until the width of about 2½ inches and removing threads from both edges until the width has been reduced to 2 inches. The test length should be 8 inches between the jaws and so enough extra length must be allowed for gripping in the jaws. If the fabric is difficult to fray along the edges, then the fabric is carefully cut to 2 in width. When the load is applied to the strip of fabric, the crimp in the direction of loading is gradually reduced and the crimp in the transverse threads increases, a process known as "crimp interchange". Occurs. The visible effect of this is the "waisting" of the specimen.

Contraction in width is greatest in the middle and decreases towards the jaws. Fig. 1(j)

In the region of the jaws, the stresses in the specimen are high and can be a cause of jaw breaks. The effect of transverse threads is very important on the strip strength. By calculating the ratio of

$$\frac{\text{strip strength per thread}}{\text{Single thread strength}}$$

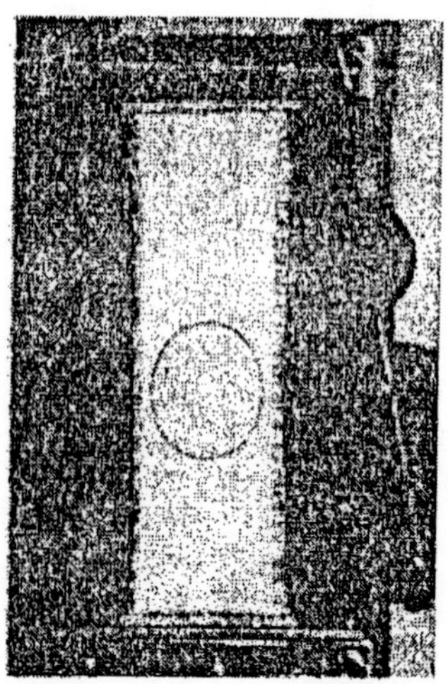

Fig. 1 (j) : Woven fabric tested in strip form.

We can get the result that is higher than unity. This indicates that the transverse threads have same form of binding effect on the longitudinal threads rendering increase in the strength. This effect is referred as fabric assistance by same authorities. This ratio of increase in strength sometimes may be 1.8 where the number of transverse threads per inch is really high.

Fig. 1 (k) : Knitted fabric tested in strip form.

The grab test—This test is conducted by grabbing the specimen between the jaws on both sides which reinforces the stresses on the fabric. A test made at the Shirly Institute

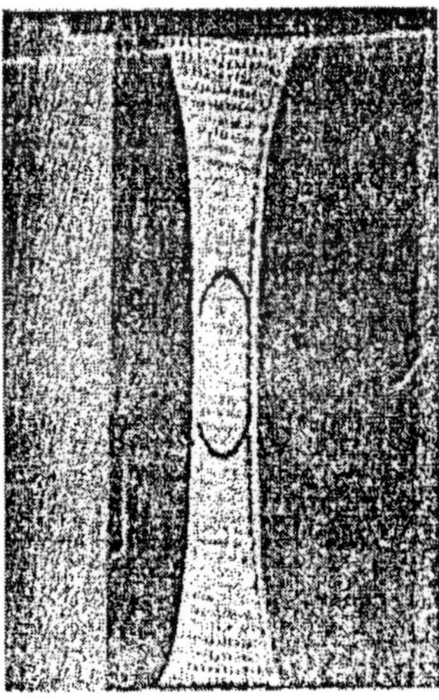

Fig. (l) : Specima of the grab test.

on a wide range of fabrics shows that the ratio grab strength tensile strength per inch for a 2 inch strip varied from 1.0 to 2.0. The ratio was found to be correlated with the breaking extension of fabric in strip test. The relation was approximately.

$$\frac{\text{Grab strength}}{\text{Tensile strength per inch}} = 1 + \frac{\text{breaking extension \%}}{40}$$

Some other factors than the extension are involved but the relation with breaking extension takes care of a large part of the variation in the grab/tensile ratio.

The Tearing Strength—A fabric which tears easily is regarded as an inferior product except where an easy tear is essential, i.e. bandages and adhesive tapes. The utility of a torn article is reduced, except that it is patched up and may be used for less important job and at worst, the article is scrapped.

Conclusion of the Resistance of Fabric to Tearing-Test

(1) Threads break singly or in small group during the tear, therefore single-thread strength of the component yarn is of great improtance.

(2) Where the fabric characteristic allow, the threads group closer together under the forces of the tearing agency. This grouping of threads is made easier if the yarns are smooth and can step over each other.

(3) Effect of weaves, Twill or 2 x 2 matt structure allows the threads to group better than plain weave. Twells and matt weave exhibet better resistance to tearing than plain-weave.

(4) High-set fabric prelude thread-movement and the assistance by thread-grouping is therefore greatly reduced.

(5) Special fabric finish such as drip-dry and crease resistance may reduce the tearing strength.

Methods of Measuring Tearing Strength

Most of the time, standard tearing testing machines are used for many tearing strength tests. Whatever may be the method, or machine, the most important factor is the method of securing the specimen in the jaws and the how the recorded strength valeus are evaluated.

For many of the tearing strength tests, standard tensile strength testing machines are used. They are modified sometimes slightly. Some authorities point out that many tears in practice occur suddenly and that a tearing test should be made at high-speed rather than at normal slow transvorse speed of tensile tests. The different methods of testing tearing strength are this not complete sentence.

The Tonque Tear Test—In this test the specimen is prepared and mounted in between the jaws of the tensile strength tester, where a pendulum lever machine is used. As the Pawls of the pendulum are made inoperative by this, the lever falls back from a peak position as the thread breaks. It again rises as the next thread is loaded for test. There are five tests done in each direction. The tearing strength reported is the median value of the peak load values.

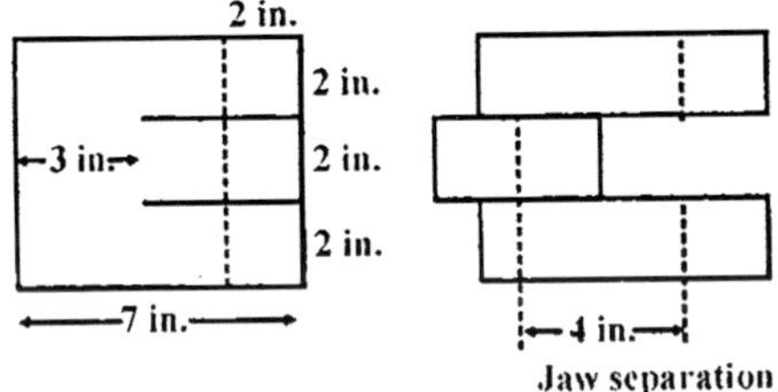

Fig. 1 (m) (i) Training Strength test specimen.

The tongue 'double rip' test—This test is similar to the tongue tear test where five tests are made in each direction. The tearing strength value is derived from the highest value during each tear and the calculation is done by taking the mean of the five highest values.

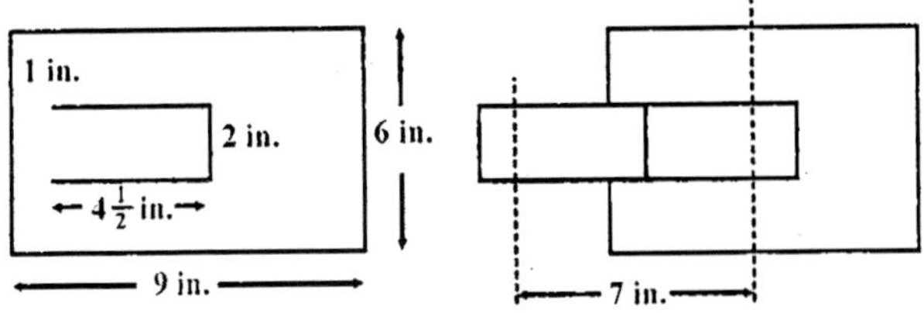

Fig. 1 (m) (ii) Tounge 'double rip' tear test.

The trapezoid tear test—This test is named so due to the shape of the test specimen. It is a standard method of The American Society for Testing Material. As in the figure, the test specimen is clamped with the jaws along the lines AB and CD, with the edge AD being held tightly and the BC edge kept in fold. the fabric tears as the jaws separate. The tear normally starts from the ¼ inch cut. The average value of five tests in each direction is taken as the tearing strength. An autographic device is useful on this testing machine.

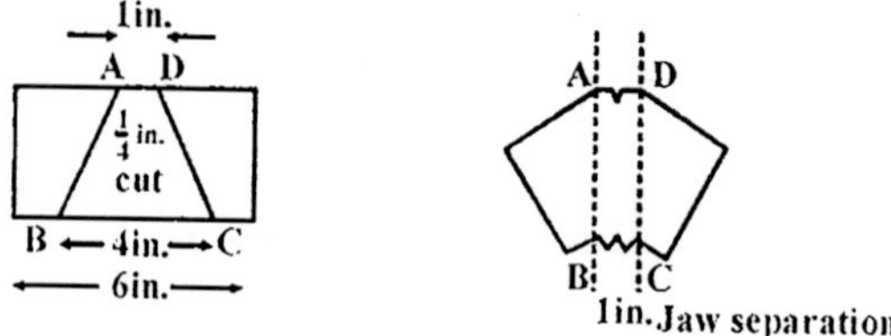

Fig. 1 (m) (iii) Trapzoid tear test.

The wing rip tear test—Preparing the specimen for the wing rip test can counteract the tendency of the tear to transfer in the direction from the direction of the applied load. For a wide range of fabrics, an angle of 55 degrees (55°) is found to be an uptimum. As the tear proceeds in this test, the point of tear remains more or less in line with the jaw centre.

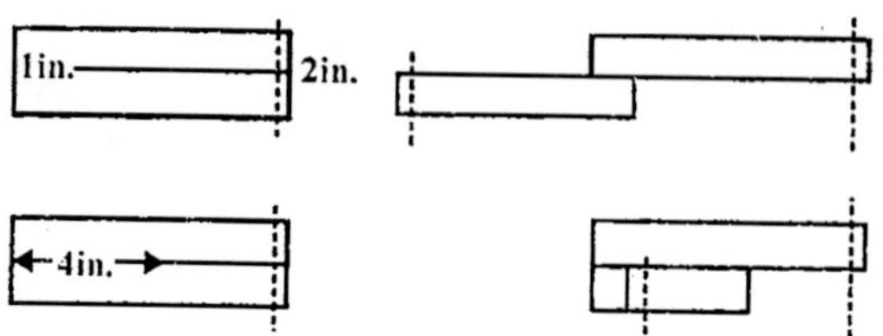

Fig. 1 (m) (iv) Ballistic 'singlerip' tear test.

The ballistic tear test—The tests conducted in the above mentioned tests are normally carried out at a slow rate of jaw separation say 4½ in/min. But tears mostly are produced accidentally and at relatively high-speeds. Ballistic test involves rapid action and thus it offers actual tearing of a fabric to a closer approximation. This test is conducted on 2 sets of specimen, five set warp way and five sets in weft way. The ballistic tester has been used to measure the tearing strength of a fabric through which a nail has been driven. This test involves tearing through 1 inch of fabric with first specimen and through 4 inch with the second set of specimen. By this, the energy absorbed in stretching the specimen tail is made possible to be cancelled out.

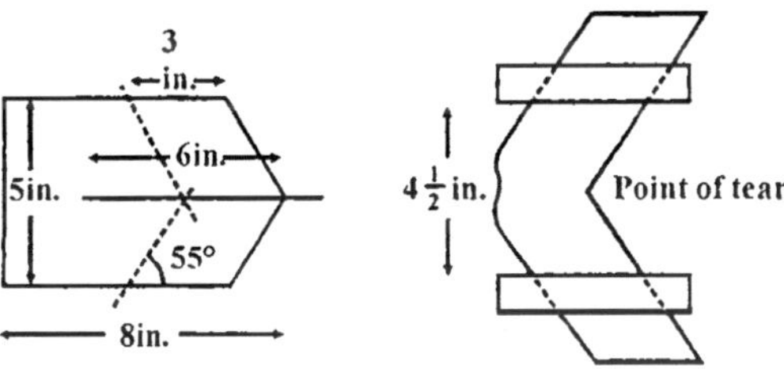

Fig. 1 (m) (v) Wing rip tear test.

Suppose W_1 = mean tearing strength of five specimens torn through 1 inch and

W_4 = mean tearing strength of five specimens torn through 4 inch

Then the Tearing strength = 1b = 1/6 ($W_4 - W_1$).

In this case the division 6 must is used in order to tear through 3 inch of fabric, having the ends of specimen being separated by a distance of 6 inch. The scale of the ballistic tester is graduated in inch pounds. The result is thus obtained in pound by dividing by inches that is 6 inches.

The textile specimens range from fine fibres to tyre cords, ropes, and plied fabrics. Besides, the methods used to mount the test specimen in the tensile testing instrument vary accordingly. Further gripping of the ends of test specimen, slip occurred during test, pressure applied by the jaws, breaking load, load extension property etc. influences the result in the tensile testing of textiles. The evaluation is to be carried out and conclusions are to be drawn accordingly.

7

IDENTIFICATION OF TEXTILE FIBRES

There are three fundamental ways of indentifying textile fibres. They are:

(1) Physical Tests,

(2) Microscopic examination,

(3) Chemical tests.

PHYSICAL TESTS

Burning of Fibres

This is a simple yet reliable test for the identification of textile fibres. For proficiency in identification by this method, however, considerable experience is needed. The student is advised to make frequent tests with known fibres until he can immediately recognise them by their burning characteristics, nature of residue and all the other clues which lead to the identification of a fibre by burning. When the student feels that he has learnt the burning characteristics of the various fibres, he should test his knowledge by arranging tests to be carried out with fibres whose origin he does not know.

The method of carrying out this test is as follows: the sample should be brought to the flame and then withdrawn. Every detail should be noted, the type of combustion, whether rapid or slow, bright or dull, the odour, the smoke (colour and density), then nature of the residue, whether the sample melts etc.

The following are the reactions of various fibres to the burning and destructive distillation tests.

Vegetable Fibres (Cotton, Linen)

These burn readily with a clear flame, and without any smell, leaving a white ash in the form of the original thread, vegatable fibres produce less smoke than other fibres.

It should be noted that cotton can be distinguished form linen by the fact that cotton fibres curl sllightly at the ends when burnt, while linen do not. When heated in a closed test tube without burning, vegetable fibres give off an acid vapour which turns blue litmus red.

Animal Fibres (Wool and Hair)

These do not born freely like cotton, but smoulder with a strong sulphurous smell. It is difficult to describe this smell, but once the student has smelt it he will be able to reconginse it on future occasions. Burn a piece of wool thread and note the smell.

Another distincts characteristic of animal fibre is the small bead which forms at the end of the burnt thread.

In a heated test tube, animal fibres give off alkaline vapours which turn red litmus blue.

Acetate, Rayon: This burns with a flame, melting with an acidic odour, leaving a hard black ball at the end of the thread. The fabric melts like sealing wax when burnt.

Mineral Fibers: These are incombustible which is of course the chief reason why these fibres are made in to fabric.

Nylon: This melts without igniting with a distinctive odour which is different from that of silk or rayon. The colour of the resultant melted mass is usually tan or brown.

Vinyon: This melts at low temperature, even when the material is at a distance from the flame.

Terylene- burns rather slowly with a sooty flame

These tests provide very good pointers towards the identity of pure fibres. Nevertheless, the student should always bear in mind the possibility of a combination yarn which would not give the clear-cut characteristics referred above.

Additional Means of Identification

It will be evident from the foregoing that the burning test is only a firs test, which identifies the group to which the fibre belongs. Other tests are necessary before it is possible to say finally the exact nature of the fabric. Very often the appearance alone will identify the fabric. For example, the burining test alone would not distinguish woollen and worsted. However, the appearance reveals the difference. Worsted yarn consists of long fibres parallel to each other and tighlty twisted to form a clean thread. Wollen yarn on the other hand is rough, and the individual fibres project all round, giving an untidy appearance. Morever, the fibres are shorter. While worsted fabric shows the weave, woollen fabric has a rough hairy face, and the weave is not so clearly seen as woolen.

Apart from the small difference revealed by the curl at the end of the thread, it is not easy to distinguish linen and cotton by burning alone. If individual threads are unravelled, however, it will be seen that the flax fibres are much longer than the cotton fibres. Flax fibres vary in length from 1 to 3 inches. The best Sea Island Cotton rarely exceeds one and half inches in length of fibre.

Another simple test which may be applied to distinguish linen from cotton fabric is the tearing test. Linen is much stronger and more difficult to tear than cotton. Moreover cotton tears with a sharp sound, while the sound of linen tearing is much duller.

Different greades of cotton may be distinguished by unravelling the threads and measuring the length of the fibre (staple). Lengths vary from 0.5 inches in the case of Chinese cotton to over 1.5 inches for Sea Island Cotton.

Amerrican and Egyptian cotton are difficult to distinguish in a finished cloth.

Laboratory Tests

Cottons are graded and classed according to their main characteristics which are:

1. Staple length.
2. Cleanliness e.g., amount of trash content.
3. Uniformity of staple.
4. Fineness and maturity.
5. Strength of the fibre.
6. Colour.
7. Ginning preparation.
8. Cellulosic degradation.

Among these staple length, uniformity, fineness, wall thickness, strength and cleanliness are capable of measurement and laboratory investigation. Colour and ginning preparation can be determined visually with some experience.

Length and Uniformity

These are measured with the help of such apparatus as the Ball's or Baer's sorters, but more accurate methods involve the use of a special instrument, the Fibrograph. This is an electronic instrument that measures the fibre length and uniformity, and indicated the results in the form of a graph, thus indicating the presence of the extent of short fibres also.

Fineness and Maturity

Fineness refers to the relative size of the cross-section of the fibre. This is also directly related to the maturity since the more mature fibres will generally be coarser than less mature fibres. For determining fineness an instrument called the Micronaire is used, which measures the resistance of the cotton fibre to air-flow. Thc lower the reading, the finer the fibre, which also indicate relative immaturity, a factor tending to cause neps during spinning and trouble in dyeing.

Trash Content

This is determined by a Shirley trash content analyser—a kind of mini opener, Microscopic examination reveals convolutions, fibre structure etc. Cellulosic degra-daion is determined by measureing the fluidity (rate of flow) of a solution of the cotton in Curprammonia.

Huntingdon Texilscope

Simple physical tests are not sufficient for the positive identification of all yarns and fabrics, particularly where there is a mixture of fibres. The Huntingdon Textilscope consi-derably widens the field of indentification.

This machine is basically a gold-leaf electroscope. A gold leaf elctroscope is fundamentally a thin piece of gold leaf, the upper part of which is attached to a support. When this foil recieves a small electrostatic charge the two arms of the foil diverge. If the charge is conducted away, the arms fall and the foil hange simply.

The textilscope depends upon the fact that animal fibres will not conduct electricity, while vegetable fiibres will. Thus, assuming that the Texilscope has recieved an electrostatic charge, so that the two ends of the gold foil are extended, if meterial made from animal fibre is placed in contact with a part of the machine connected electrically to the metal foil, nothing will happen. On the other hand, if material or yarn made of vegetable fibre is made to touch a conducting surface of the machine, the two arms of the foil will descent as the electrical charge is conducted away.

Many different tests may be made with this machine as different fibre have different conductivities. Thus for different threads and cloths, the arms of the metal foil will descent at different rates, experience and knowledge of the machine indicating the nature of the fibre. The following three main tests should be noted.

(1) Preliminary or Piece Test: This is the simple test described above, to ascertain whether they are of animal or vegetable origin. If the leaves descent, the fabric is made from vegetable origin. If the arms remain horizontal, the fabric is made from pure animal fibre. The rate at which the arms descent is timed, as this differs with different vegetable fibres and is thus an important means of identification

(2) Thread Test: Individual threads are held against the detector. If the material is a mixture of anmial and vegetable fibres, this will be revealed by unravelling the individual fibres and noting the different reaction produced by each.

(3) Oscillatory Test: For this test a single tread about two inches long is held tightly by one end while the other end is brought very close to the indication of the detector. The action of the thread is then closely observed, threads of different origin reacting in a different manner due to their different electomagnetic properties. The distance from which the thread is affected must be observed.

The action of the more common fibres is as follows:

Silk: Piece test and thread test. The indicators remain rigid. If the silk is loaded the indicators descent slowly or rapidly in accordance with the extent of the loading.

Oscillatory Test: When the tread is brought near the indicator, it will be attracted and then released.

Wool: Piece Test and thread test. Indicators remain rigid.

Oscillatory Test: No reaction. The thread is not accecpted.

Cotton: Piece Test. The indicators descend immediately.

Thread Test: The indicators descend slowly, taking about ten seconds.

Oscillatory Test: The thread is attracted and held as the indicators descent.

Linen: Piece Test. The indicators descent immediately.

Thread Test: The indicators descent slowly.

Oscillatory: Immediately attaction and oscillaration, more marked than in the case of cotton. This lasts about five seconds and the indicators descent.

Viscose Rayon: Piece Test. The indicators descent rapidly.

Thread Test: The indicators descent less rapidly.

Oscillatory Test: The thread is attacted when half an inch away from the detector, and then repelled forcefully. The attraction is slightly more than that of normally loaded silk.

Cuprammonium Rayon: Piece Test. The indicators descent fairly slowly, taking about two seconds.

Thread Test: Very little action, usually the indicators take about twenty seconds to descent.

Oscillatory Test: The tread is attracted one fourth inch from the indicator, then repelled. It may afterwards be re-attracted.

Acetate Rayon: Piece Test and Thread Test. The descent of the indicators is very slow and in fact they may remain horizontal for some time.

Oscillatory Test: Almost impereeptible attraction about one eighth of an inch from the detector.

These test must be carried out with the sample under test and the hand dry, as of course, the presence of moisture alters their conductivity.

Identification by use of the Microscope

The Microscope is a valuable means of identification of fibres. All the common fibres have a distinctive appearance under the microscope. Moreover, by using an ocular micrometer disc the comparative diameter of fibres can be measured. Such small measurements are computed in thousandths of a millimeter, the symbol for which is u. A compound microscope is needed for work of this kind, as of course a simple microscope has insufficient magnification.

As a guide to students we give below and in the attached diagrams the characteristics of the more common fibres as seen under the microscope. Students who have access to a textile laboratory are advised to inspect common fibres for themselves.

Cotton

The striking feature of cotton fibres is the frequent twists along the length of the fibre. The general form is that of flattened tube, with thickened edges and a wide lumen (note-lumen is the hollow space between the cell walls).

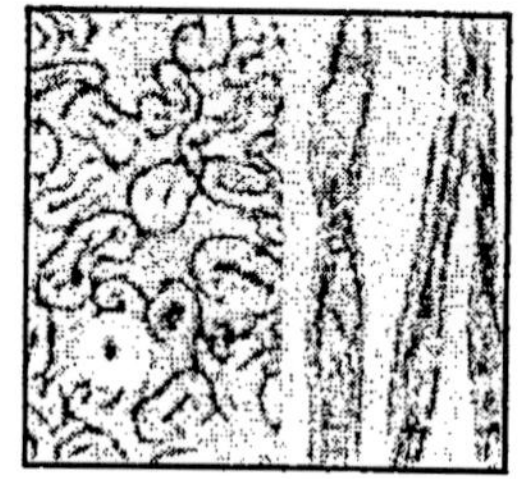

Fig. (1) cotton.

Mercerised Cotton

In well mercerised cotton the twisted are almost absent. The fibres are well rounded with a narrow lumen. The mark of an occasional twist helps to identify mercerised cotton from other smooth fibres. (Fig.2)

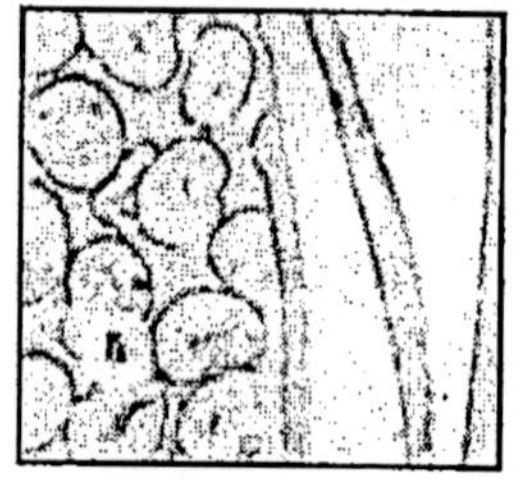

Fig. 2 : Mercerised Cotton.

Flax

Despite the similarity of cotton and flax under other tests, under the microscope there can be no possible confusion. Flax is absolutely distinct, the fibres being cylindrical with knots like bamboo at regular intervals. The shape of the corss-section of linen is polygonal with the lumen as a dot Figure (3) fax.

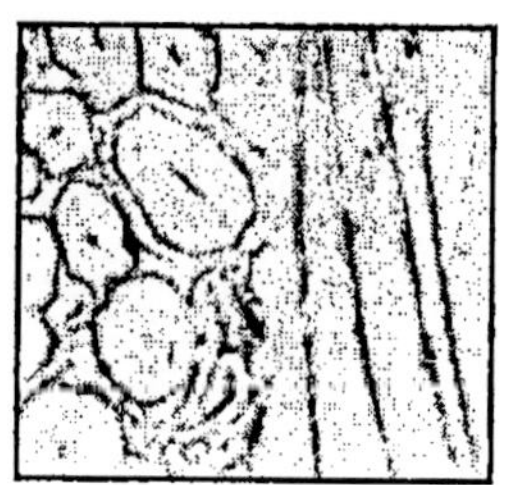

Fig. 3 : Flax

Wool

Wool has a distinctive scaly appearance. The fibre is covered with horny overlapping scales. Note the rounded form of the cross-section (Fig.4)

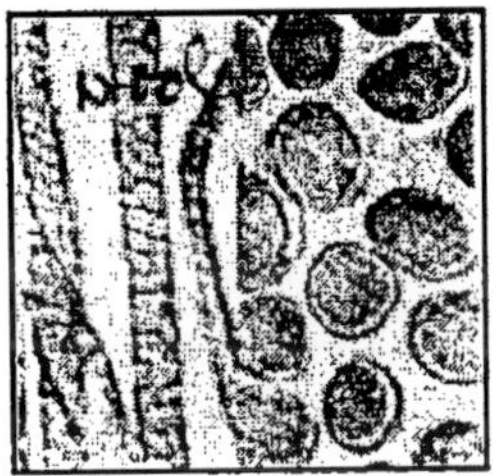

Fig. (4) : Wool.

Silk

This appears as smooth, structureless and transparent filament. Unlike cotton, falx and wool it is not of organised structure.(Fig.5).

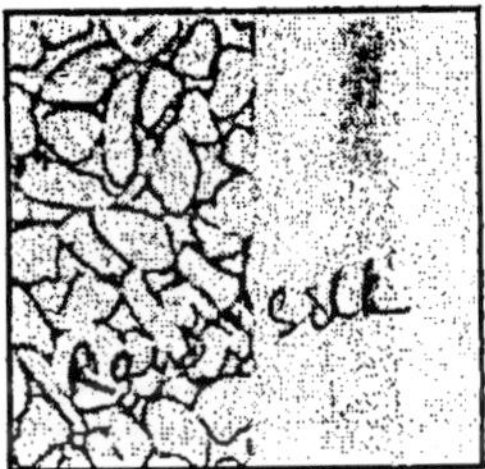

Fig. (5) Raw Silk.

Tussar silk

The characteristics are broad filament, flattenced ends and irregular markings. The diameter is irregular.(Fig.6).

Fig. (6) Tussar silk

Rayons

Longitudinal examination reveals little, and the cross-sections are important. Viscose rayon has a serrated margin in cross section (Fig.7).

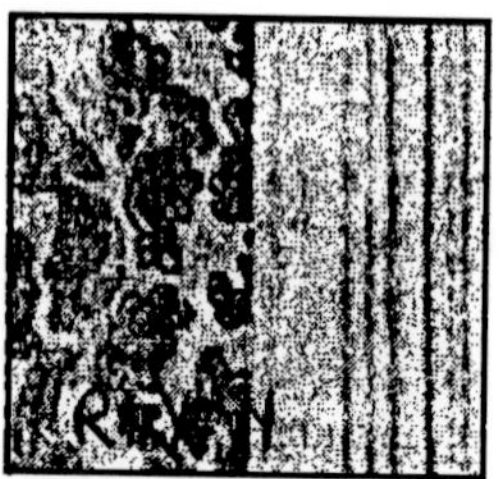

Fig. (7) Viscose Rayon.

Acetate rayon has a lobar cross section

Cuprammonium rayon is circular in cross section. (Fig.8).

Fig. (8) Cuprammonium Rayon.

Nylon

This is structureless and cylindrical in appearance, while in cross section it is smoothly round. Aralac fibres have almost the same characteristics under the microscope as nylon.

Fig. (10) : Nylon Fibres.

Terylene

Cylindrical and smooth appearance with a round cross section

Fig. (11) : Terylene.

Vinyon

The longitudinal appearance is cylindrical though in cross section it has the characteristic shape of an oval narrowing at the middle.

Fig.

Microchemical tests are sometimes carried out by adding one or two drops of certain liquids and watching the reaction under the microscope. Permanent photographic records of textile fibres under the microscope should be kept for comparative purposes.

DIAGRAM OF VIEWS OF FIBRES UNDER MICROSCOPE

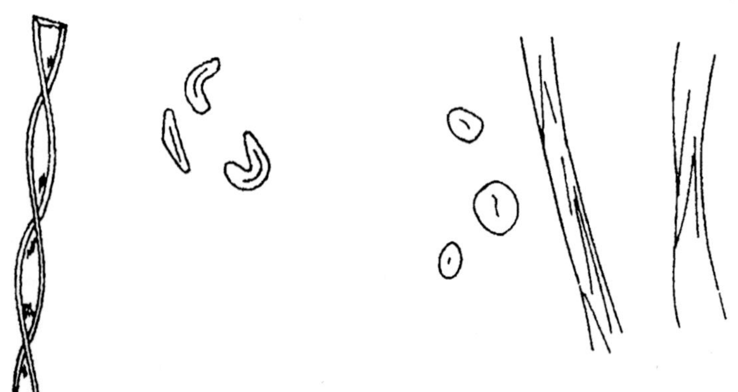

Fig. 1 : Cotton fibres. Note collapsed ribbon form. The lumen is visible in cross-section.

Fig. 2. Mercerised cotton. Note rounded form. Faint traces of twist are visible

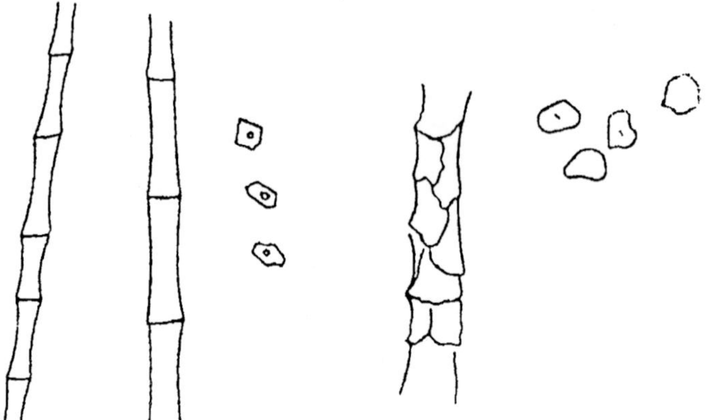

Fig. 3 : Note bamboo shape. Cross-section is irregular polygon.

Fig. 4 : Note the overl soales and round cross-section.

Fig. 5 : Raw silk

Fig. 6 : Silk fibres (degummed)

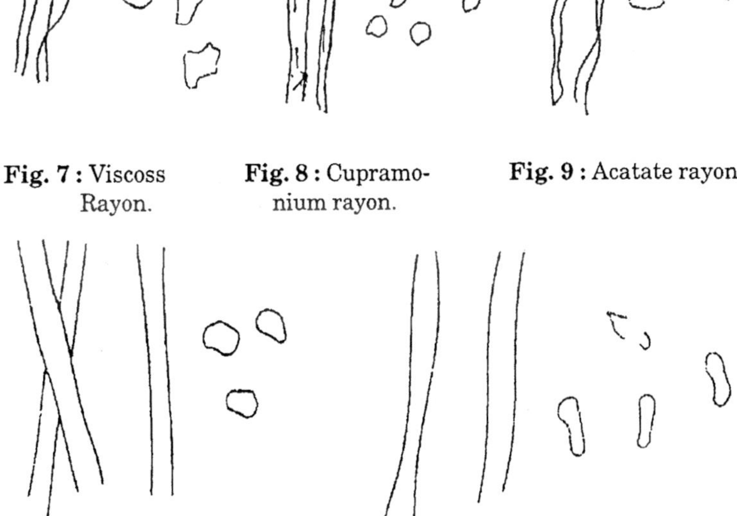

Fig. 7 : Viscoss Rayon.

Fig. 8 : Cupramo-nium rayon.

Fig. 9 : Acatate rayon

Fig. 10 : Nylon fibres

Fig. 11 : Vinyon fibres.

CHEMICAL TESTS FOR TIEXTILE FIBRES

Cotton

When a small portion of the sample is immersed in strong hydrocloric acid it is completely decomposed. When the resultant lignid is evaporated, the residue is a powder. Cotton is also decomposed by a strong solution of nitric acid. Iodine solution stains cotton blue.

Mercerised Cotton

We have just mentioned that iodine solution stains cotton blue. An extension of this test may be used to differentiate mercerised from unmercerised cotton. After the sample of cotton cloth has been stained blue, it should be washed. If the colour remains, the cloth has been mercerised. The blue colour washes out of unmercerised cotton.

Silk

Hot caustic soda or hot caustic potash dissolves silk. Cold hydrochloric acid, nitric acid or sulphuric acid will dissolve silk. It should be noted that the process is a rapid one. Tussor silk may be distinguished from ordinary silk by the fact that while it dissolves in the aforementioned acids, the process is a slow one, taking as long as an hour in some cases.

Wool

Wool is entirely dissolved by a 5% solution of caustic soda. If a few drops of lead acetate are added to the resulatant liquid a black precipitate forms. Wool swells when treated with hydrochloric acid. Nitric acid truns wool yellow.

Jute, Hemp and Flax

These may be tested by immersing the sample in iodine

solution for three to four minutes. The sample is then washed in dillute hydrochloric acid and finally wased in water to remove all trace of Iodine. Reactions are as follows:

Jute—the sample turns to orange colour.

Hemp—the sample is unaffected.

Flax—the sample turns blue.

Wool and cotton mixtures may be detected as follows.

The sample is placed in a 5% caustic soda solution for 30 minutes. Any wool present is entirely dissolved while cotton is unaffected. If lead acetate is added to the resultant solution, a black precipitate forms if wool is present, while a white precipitate indicates cotton alone.

Where an actual quantity test is required to ascertain the exact proportions of cotton and wool in a fabric, the procedure is as follows. First the sample is weighed very carefully. It is next placed in a strong solution of hydrochloric acid which has the effect of dissolving out the cotton. The fabric is then dried, washed in pure water and afterwards boiled in pure water for ten minutes. The next step is to dry the sample in moderate heat to allow it to regain natural moisture. The residue is weighed and the difference in weight gives the percentage of cotton removed.

A simple physical test for cotton and liner mixtures is to place a drop of ink on the material. If the fabric is pure linen, the ink will spread rapidly in all directions. If the fabric is a mixture, the ink will spread rapidly along the linen thread and much more slowly along the cotton threads. If the material is pure cotton, the ink will spread slowly in all directions.

Chemical tests for linen and coton mixtures are as follows. Soak the sample in methylene blue solution and then rinse it in water. The linen threads keep the blue colour of the methylene blue, while the cotton threads lose their colour on rinsing. Another test is to soak the sample in cold caustic potash. Both cotton and linen contract and twist, but the cotton remians grey, while the linen turns yellow.

To distinguish silk and rayon, place the sample in a cold saturated solution of chronic acid mixed with the same volume of water, raise to the boil and continue boiling for one minute,. Silk will be completely dissolved while rayon is unchanged. Care must be taken that the period of actual boiling does not exceed one minute.

To distinguish various rayons place the sample in a solution of sulphuric acid and iodine in equal parts.

Viscose rayon truns dark blue. Cuprammonium turns light blue.

Acetate rayon turns yellow.

Vinyon and nylon may be distinguished by the fact that an 80 % acetone solution dissolves vinyon but not nylon, while on the other hand an 80 % sulphuric acid solution dissolves nylon but not vinyon.

Solvent tests for synthetics

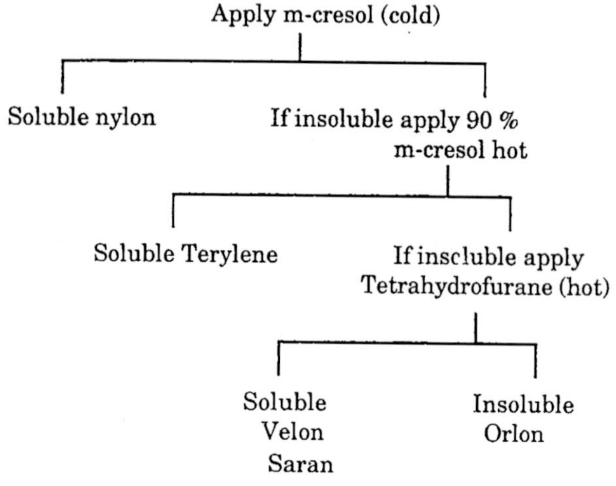

8

TESTING OF YARN

FIBROUS ADULTERATION OF YARN

The student must learn to detect the presence and extent of adulteration in yarns and fabrics. This problem does not arise in the case of yarns and fabrics made form pure fibres, as the nature of the yarn is revealed by its name. There are however, certain yarns whose names do not indicate the nature of their constituents. They have extensive use and their names should be known.

Angola Yarn

This yarn has nothing with the Angola Goat. It is merly a trade name for a mixed yarn of wool and cotton. The wool comes from old hard wool rags, old suits, etc., and is extracted by carding and felting the material. These rags are processed by special machinery to produce a fluffy mass which cannot be spun. When cotton is mixed with it, the mass becomes closely tangled with the cotton fibres and can be spun into yarn. The cotton gives strength the wool softness, brightness and elasticity. There are various types of Angola yarn and they can be distinguished by the percentage of wool used in the manufacture. Angola Yarn may contain form 15 % to 70% wool. The exact percentage may be assessed by a quantitative test as explained in the previous lesson.

Shoddy Yarn

This is an all wool yarn made from various unfelted soft wastes. Though of inferior quality it is very cheap and useful, as wool garments of all types may be made from it.

Cotton Shoddy Yarn

This is made from cop bottoms and other hard waste. It is of poor quality but has many uses including the manufacture of cheap cotton blankets, candlewicks, cotton waste goods, etc.

Mungo Yarn

This is an inferior type of all wool shoddy. It is made from hard woollen cloths, carded and milled rags. The tearing process weakens the wool fibres, as the covering scales which hold wool fibres together are stripped off. Though both mungo and shoddy are processed identically, the yarns are entirely different in character.

Extract Wool Yarn

This is made from wool and vegetable fibre mixtures by dissolving out the vegetable fibre with sulpuric acid. The remaining wool or animal fibre is collected and spun. It is of very poor quality and inferior to mungo yarn.

Woolen Yarn

This is useful yarn made from the short fibres thrown out by the worsted combing machine. Though it has no staple, the yarn is soft and elastic with good felting properties.

Spun Silk Yarn

As its name reveals, this is made from various kinds of waste silk which are spun together. This is serviceable though not very brilliant owing to its short staple.

The students should endeavour to obtian sample of all the mixtures referred to here in order to acquaint themself with their appearance and feel. This will help them greatly in subsequent analysis. In fact, textile students should maintain a textile scrap book, into which they insert samples of all common material with definition, classificatioin of count, weave, fibre, etc. If properly maintained and catalogued, this book will be of great value to them.

GENERAL EXAMINATION OF YARN

All fibres except silk must be spun into yarn before they can be made into cloth. Spinning may be described as the process of laying the fibres paralled to each other and then twisting them together. The qualities necessary for spinning are tensile strength, elasticity and spinning power. Spinning power comes from the curl or natural twist present in the fibres, enabling then to hold together during the process of spinning. Spinning power is largely influenced by the length of the fibre and the finess and evenness of its diameter.

Yarns are spun in such a way that there is a definite relationship between length and weight of yarn. This relationship is known as count, which will be fully explained later. In order to maintain the count constant the diameter of the yarn should be regular, and this is dependent on the evenness of the individual fibres. The twist of the yarn, i.e., the number of complete turns in a given length-must also be constant, and this is dependent upon uniformity of the diameter of the yarn.

As you know, there are two distinct threads in a fabric, the warp and the weft. Each has a distinct function and hence a different yarn is required in general for each. The warp thread form the skeleton of the fabric undergoing the strain of manufacture, and therefore, the primary requirements are tensile strength and elasticity. On the other hand, weft threads from the main body of the fabric and thus the first requirement is filling power rather than strength, though of course some strength is needed. Warp yarns are often artificially strengthened by sizing during the manufacturing process.

PHYSICAL PROPERTIES OF YARN

The following are the characteristics to be assessed and tested when judging yarn:

(1) Percentage of moistrue
(2) Count
(3) Eveness and regularity
(4) Elasticity
(5) Strength
(6) Twist (per inch or other length)
(7) Colour

(1) Moisture: Cotton yarn fresh from the spindle is dry and light. The fibres project all round the thread. Moisture must be added befor the cotton can be satisfactorily spun, but it should be sufficient to enable the cotton to be spun and no more. Cotton can absorb as much as 20% by weight of water feeling damp. However, cotton is bought and sold by weight at this state and hence, the need has arisen of standardising the percentage of moisture added by conditioning. As the natural constituent of cotton is 8.5 % regain, this amount of moisture should be added before the spinning process. This makes the cotton sufficiently supple for spinning and levels out the projecting fibres. There is a further loss of moisture during spinning and this should be replaced by conditioning though more than 8.5% is excessive. Excess moisture can always be detected by a simple test with a moisture testing oven.

The moisture testing oven consists of cage hung from a balance and subjected to a blast of hot air. The sample under test is placed in the cage and weighed. It is then dried in a steam of hot air and when it is completely dry it is weighed again. The temperature of the oven is maintained at 100°C to 105°C, and the sample is assumed to be completely dry when there is no change of weight for ten minutes. The difference in weight of the sample before and after the test represents the weight of the moisture originally contained. This can be calculated as a percentage.

(2) Count: The relatioship between the length and

weight of yarn is known as count. Different methods of computation of count are used for different fibres. Count is expressed as a number 40s-40 count.

Cotton count is the number of hanks of 840 yards length which weighs one pound. Thus, if 40 hanks of 840 yards weigh exactly one pound the yarn is said to be 40 count or 40s. This system is also used for spun silk yarns. When two single threads are combined to make a yarn of double thickness the yarn is said to be doubled or folded. The count is expressed on the basis of the original yarn. For exmple, if 60s yarn is double the result in a folded yarn equivalent to 30s. This is expressed however as two-fold 60s or 2' 60s.

For worsted, alpaca, camel-hair and similar yarns the length of hank used as a basis is 560 yards. There are various system in use for woolen yarn. The system most nenerally used in England is based on the number of hanks of 256 yards length which make up pound. There are two systems in use in America, the Cut system and the Run system. A cut is 300 yards and a run is 1,600 yards, the weight unit being one pound.

Cotton and woollen yarns of very low count, in other words the very thick yarns, express the count on a yard per ounce basis.

Linen is based on a length of 300 yards known as a Lea. The number of leas which go to make a one pound weight is the count of linen. Fine jute and Ramie yarns are also numbered on this system.

The foregoing systems are based on a fixed weight, the variable factor being the length of the piece of yarn. Some fibres however use another system whereby the weight is variable although the length of the hank is fixed. Typical of such systems, are those used for calculating the count of silk yarn coarses, Jute and Hemp counts are based on the weight in lbs. Of one spindle i.e., a length of 14,400 yds.of yarn.

There are two system used for the assessment of Silk yarn. These are the gram system and the denier system. In the case of the gram system the count is the weight in grams of one thoudand yards of yarn. The denier system is more popular and in common use. In practice, the count by the denier system is the weight in grams of 9,000 metres of silk yarn. This system is also used for determining the counts of filament manmade yarns of regenerated as well as synthetic origin. Spun yarns are however based on the spinning system used viz., cotton, worsted etc. In addition to the systems given above various attempts have been made to introduce universal counting systems with varying degress of success. The main ones are as follows:

Metric Counts: This is an indirect system similar to cotton, worsted etc. However, the "Hank" length is changed to one km. i.e., 1000 metre and the weight to one kgm i.e., 100 grms instead of 1b.

Grex Count: This is a direct system. It is the weight in grams of 10,000 metres length of yarn.

Tex System: This system has been introudced by the Textile Committee of the International Organisation for Standardisation and recommended for adoption as a universal yarn numbering system. This is a fixed length (one thousand metres) and variable weight (grams) system i.e., a direct method. It can be easily seen that the Tex standard is 1/9 of the Denier and 1/10 of the Grex standard.

Testing of Count-Long Lengths

Counts of yarn can be estimated from either long or short lengths of yarn. If long lengths are available the counts are calculated from lengths of yarn in lea or hank from. In the case of cotton the lea is 120 yards and the hank 840 yards. The length of yarn is first measured on a wrap reel, which is a winding reel with a recording dial which shows the length of yarn wound. When a definite length of yarn has been reeled, it is removed and weighed in a sensitive balance. The count is calculated as follows:

$$\text{Count of cotton yarn} = \frac{\text{Length weighed}}{\text{Weight in grains}} \times \frac{7{,}000}{840}$$

It should be noted that there are 7,000 grains to a pound.

There are available certain specially manfactured balances which record the actual count of a given length of yarn when a hank is placed on the hook provided. These may be used for different fibres. They are often delicate machines and require careful handling.

TESTING OF COUNT—SHORT LENGTHS

To measure the count of cloth, a number of threads, both warp and weft should be selected form places some distance aprat in the fabric. The thread should be carefully straightened out and all curl removed. If the fabric has been starched or weighted, this additional matter should be removed. The total length of all the threads is measured and then they are all weighed together. Given the length and weight, count can be calculated. It will be appreciated that such work calls for a high degree of accuracy and a number of tests and checks. There are available certain machines and balances which simplify the calculation of count form small pieces.

For ordinary calculations, especially if the sample of yarn is a long length, the question of moisture requires no special considerations. In cases where the sample is small, or where there is a dispute between two parties concerning the exact count, it is necessary to allow for the moisture content of the yarn.

COUNT CALCULATIONS

A— Indirect or reciprocal systems viz., those based on fixed weight and variable length.

The formula for calculation of counts of cotton yarn has been given earlier. However, a general method of calculating counts applicable to all indirect systems including those for cotton, worsted, woolen, linen, etc., is given below. This is useful in calculation other determinants if required as shown in the:

$$\text{General formula}\ \frac{L}{W \times C \times S} = 1$$

where L = length of yarn in yds.

W = weight of the yarn in 1bs.

C = count of yarn.

S = standard number (i.e., hank length in the system).

It is thus seen that when L and W are known, only the standard length required to be changed viz., 840 for cotton, 560 for worsted, 300 for lined etc., for finding out the counts in the respective system. Moreover, if the counts are known either the weight or length can be calculated if one of them is given.

Example: find the weight in lbs. of 80,640 yds of a 36s worsted yarn.

Substituting in the General formula

$$\frac{80.640}{36 \times 560} = W = 4\ \text{lbs.}$$

B-Direct systems viz., fixed length and variable weight

$$\text{General Formula}\ \frac{S \times W}{L \times C} = 1$$

Where S = standard length in metres (corresponding to hank length)

W = Weight in grams

L = Length in metres

C = Count

N.B. substitute yds and 1bs for metres and grams in case of Jute and Hemp or where applicable.

Example: What is the Denier count of a yarn of 450 metres length which weigh 10 grams.

$$\text{Substituting in the formula}\ \frac{9000 \times 10}{450} = C = 200\ \text{Den}$$

EQUIVALENT COUNTS

Between Indirect Systems

Problems often arise of converting one indirect count into another e.g., cotton to Worsted or vice versa, worsted to woolen etc. In such cases the general formula is

$$C = \frac{R \times S}{G}$$

Where C = The system to which the conversion is to be made.

R = Given count which has to be converted.
S = Standard for the given count.
G = Standard for the new system.

Example: What is the equivalent in cotton counts of a 60s worsted yarn?

Substituting in the formula

$$C = \frac{60 \times 560}{840} = 40\text{s cotton}$$

Indirect to Direct System

Cotton to Denier counts and vice versa.
Formula: Cotton count x denier count = 5315
Worsted to Denier counts and vice versa
Formula : Worsted count x Denier count = 7973
Formula : Metric count x Denier count = 9000
Tex conversion : Tex count = 9 x Denier count

Tex count = 10 x Grex count.

Examination of Yarn for Quality, Evenness and Regularity:

Before yarn is woven into cloth it should be inspected for quality, evenness and regularity. Uneven patches, knots, and neps etc., will mar the appearance of the finished fabric. The examination is carried out by drawing yarn from one or more cops or bobbins and spreading, it over a sheet of cardboard with a dead black surface, leaving sufficient space between the threads for defects to be seen.

Tensile Strength and Elasticity:

Tensile strength of a thread of fabric is measured as the weight required to break it under tension or in other words under a steady pull as distinct from a sudden jerk. Elasticity may be defined as the power of a thread or fabric to recover its original length when the tension or stretching force is removed. A number of factors affect tensile strength and elasticity and among them are:

1. Length of fibre.
2. Fineness of fibre.
3. Twist in fibre.
4. Wax content of the fibre.
5. Hygroscopicity (ability to absorb moisture) of the fibre.
6. Count of yarn.
7. Twist.
8. Moisture content of the yarn while under test.

One machine can be used to test both tensile strength and elasticity. There are a number of different/machines, either for testing a complete hank or lea or a single thread. The lea testing machine gives only a rough and ready test, as of course the breaking point represents the average of the whole length. The basic principle underlying these machines is that a length of the thread under test is tautened steadily usually by a small electric motor, until the thread or lea breaks. The acutal stretch before the thread breaks is measured on a scale and this gives the elasticity. Another scale indicates the power expressed in pounds weight which is required to break the sample. There are also manually operated machines for carrying out tests of elesticity and tensile strength, but these require very skilled operation to ensure that the pull on the thread or lea is uniform and does not vary during the test. Some single thread machines use hydradulic power.

When carrying out these tests, care is needed to ensure that the thread is not excessively twisted as a result or careless reeling, as this will increase the strength of the yarn beyond its true normal strength. In the case of the lea test, the lea must be very carefully wound to ensure that all rounds of the yarn are of the same length. If the tension is uneven during winding, the lea will not take up the strain as a whole, and only part will receive the strain at first. This will lead to inaccurate results. The machine itself should be tested before the test by the use of actual weights to check that the dial is reading accurately. It should be noted that cops taken directly from the conditioning cellar will give different results from those which have been some time in the testing room due to variation in moisture content.

Elasticity is expressed as a percentage. This is the percentage of the original length which the thread or lea stretches before breaking. Thus, if a lea of 20 inches long stretches 5 inches, then its elasticity is 25%. It should be noted that when a lea is tested some thread always break before others, and so the elasticity test is not 100 % accurate. A single thread gives an exact test for that particular length of thread, and if the thread is uniform, for the whole lot under test. Of course, in practice a number of tests are made for each lot under test.

Twist in Yarn

This may be defined as the number and direction of turns given to a particular length of the yarn during spinning. There are two directions, the right-hand/or Z twist and the left-hand or S Twist. Right-hand twist is known as twistway and this twist usually denotes warp yarn. Left-hand twist usually denotes weft yarn and is known as weftway. To ascertain the twist hold the thread between the thumb and forefinger of the left hand. If it untwists on turning outward with the right hand it is twistway. If it untwists when the right hand is turned inwards it is weftway.

Table : Yarn Properties of Different Fibres

Types of Yarn	Sectional Shape	Specific Gravity	Dry Tenacity	Extension at break %	Ratio of wat Tenacity Dry Tenacity	Moisture Regain	Modulus of Elasticity g. p. d.
Terylene (or Dacron) Filament Yarn	Circular	1.38	4.5-5.5	27-17	100	0.4	110
Nylon Filament Yarn	Circulular	1.14	4.5-5.5	25-20	85-90	4.2	24
Terylene Staple	Circulular	1.38	3.5-4.0	40-30	100	0.4	50-55
Orlon (42)	Irregular dog-bone	1.16	2.2-2.6	24-28	80-82	1.5	–
Viscose Rayon	Serrated	1.52	2.1	21	44-54	11.00	65
Acetate Rayon	Trefoil	1.33	1.3	29	60-65	6.0	31
Cotton	Bean-shaped	1.52	3.5	7.3	110-130	8.5	55
Wool	Circullar	1.32	1.1	38	78-90	14.5-16.5	28

The relationship between twist and count for cotton and worsted is as follows:

Cotton : Twist in warp $= \sqrt{\text{count} \times 3.75}$

Twist in weft $= \sqrt{\text{count} \times 3.25}$

Worsted : Twist in warp $= \sqrt{\text{count} \times 3.58}$

Twist in weft $= \sqrt{\text{count} \times 1.29}$

The twist of yarn can be found by the use of a simple mahcine which unwinds the thread and records on a dial the number of turns needed to unwind the thread completely. The number of turns made to unwind the thread divided by the length of the piece under test in inches gives the number of twists per inch. Care should be taken to ensure that the threads have not started to twist beyond the fully unwound point. This can be checked quite simply by passing a pin through the length of the thread. There is less likelihood of any mistake being made on this score if a fairly short piece of thread is tested. It must be remembered however that twist varies from inch to inch in a thread, and hence a longer piece gives a more accurate average figure.

Gassed Yarns

These are yarns which have been passed through a gas flame or over a hot singeing plate to burn off projecting fibres and to give the thread a smooth, round appearance. Gassing can be detected under a strong magnifying glass or a low-powered microscope.

Testing of Cotton Waste Yarns

Experience is an important factor, and an expert can judge cotton waste by sight and feel alone. Cotton waste is generally tested by finding the weight and calculating the percentage of different wastes mixed in the sample under test. The percentage of different constituents will indicate which is the best waste for a particular purpose. Cotton waste can also be separated into its constituents parts mechanically.

9

TESTING OF WOVEN FABRICS

Woven Fabrics

The student will at this stage have a clear idea of what constitutes a woven fabric, with the warp threads running the length of the fabric and the weft or pick or filling crossing at right angles forming the selvedges at the edges.

The following particulars must be obtained before a satisfactory analysis of a woven fabric can be made:

1. Width and length.
2. Determination of warp and weft.
3. Number of ends per inch.
4. Number of picks per inch.
5. Material or materials used for weft threads and their colours.
6. Counts of warp threads.
7. Counts of weft threads.
8. Weave or design of the fabric.
9. Particulars of the selvedge.
10. Finish of the fabric.
11. Weight of the cloth per unit area.
12. Shrinkage if any of warp and weft.
13. Tensile strength of the fabric-warp way and weft way.

Width And Length

It should first been noted that cloth contracts after leaving the loom. The actual contraction varies with the weave of the cloth and the tension exerted upon it while it is within the loom. Another factor is humidity which may lead to shrinkage in width.

Even allowing for these variations, differences frequently arise between manufacturers and their clients concerning the length and width of a particular piece of cloth. The manufacturer may take the view that the cloth has the correct number of ends, and on these grounds maintain that the width is correct. But the customer is not interested in such matters. His measure is the yardstick, and hence the munufacturer must also concern himself with actual measurements.

The width of cloth should be measured as follows: First, lay the material out carefully on a flat surface, and ensure that all creases are removed, but that the material is not under tension anywhere. The width should be measured at several places and an average is taken. Heavier cloths measure well, but great care is needed in measuring mulls, muslins, voils and similar light clothes as they are very elastic.

The length of cloth is affected by humidity and the natural elasticity of the material. In the case of cotton goods there is an appreciable shrinkage a few days after leaving the loom due to the release of tension on the warp threads and the natural elasticity or cotton.

Various yard sticks are available, but the most suitable type for most purposes should be flat with metal tipped ends, and inches marked on both sides. The metal ends should be tapered to prevent the cloth from wrapping round them.

Determination of Warp and Weft

There are a number of pointers to indicate which is the warp and which is the weft in a given fabric. These are as follows:

(1) Selvedge reveals the warp and weft unmistakably.

(2) Read marks indicate warp. Reed marks are regular groupings of the threads in threes or fours.
(3) Generally warp thread are spun twist-way and weft threads are spun weft-way.
(4) As a rule, the harder twisted of the two sets of threads is warp, and the soft spun is weft.
(5) Regularity in the thread indicates warp, irregularity weft.
(6) If of the two sets of threads one has been sized, that one will be warp, as weft thread is very rerely sized. This cannot be seen in a finished fabric as the thread sizing is wased out of the fabric.
(7) The tension of the loom makes the warp threads straighter in the cloth than the weft.
8) Coloured strips are more usual in the warp than the weft.
(9) Variation in the counts of the thread in one set indicates warp.
(10) Generally check looms put picks in pairs. Odd numbers of threads in one colour in a check pattern can be taken as warp. The longer checking pattern is also generally warp way.
(11) Folded yarn is more commonly used in warp than weft.
(12) If in a cotton-wool mixture fabric, the cotton is used all one way, the cotton is warp.
(13) In a striped cloth, crammed yarn one way indicates warp.
(14) In general, if the threads are equal in all other respects, the stronger of the two will usually be the warp.

The Face Side of Cloth

Determination of the face side of fancy cotton cloths rarely presents any difficulty. The following should be noted:

(1) Satins have the warp on top.
(2) Sateens have the weft on top.
(3) Coloured clothes are generally brighter on the face side of the cloth.
(4) Cords are more pronounced on the face.
(5) Dobby patterns and brocades are more pronounced on the face.
6) Finishes such as schreinerising, printing etc. reveal the face side of a cloth.

DETERMINATION OF THREADS PER INCH

The number of thread per inch may be determined by one of two methods. These are:

(1) Counting the ends and picks under a magnifying glass.
(2) Cutting one square inch from the sample and counting the threads one by one.

There are various types of coutning glasses of varying sizes. A useful glass for everyday use is the one inch square glass with double glass lens. This shows corners and sides clearly thus ensuring an accurate count. The use of counting glasses smallar than one inch opening is not recommended. The number of threads seen under a quarter inch glass may vary. One person may count 16 threads and another 15, as one man may count all threads visible under the glass while another may start with the first complete thread visible. The latter course is safer, if the use of a small aparture glass is unavoidable.

A useful glass is the American pattent 26473, 10/3/1910, which leaves the hands free to work with glass and cloth. Moreover it incorporates a triangular scale, each side of which may be aligned to the cloth to simplify counting and measurement. This scale also acts as a light reflector.

The visual counting of cloth is very difficult in the case of some finished cloths, especially those of dark colour.

In such cases a reflector is needed. This may take the form of a box, or even a complete table with a groundglass top, illuminated from beneath by a powerful electric light. This shines clearly through the cloth, showing the weave clearly, and facilitating the counting of threads.

The counting of threads in each direction–warp-way and weft-way should be done at three or more places. The readings should be made several inches apart as permitted by the sample under test. Warp readings should not be taken near the selvedge.

In the case of certain fabrics which have a complicated weave, it is sometimes difficult to count the number of threads accurately with a counting glass. The only really reliable way to make a count in such cases is to cut an exact inch and separate the threads, counting them one by one. First separate a pair of needle pointed dividers exactly one inch by placing the needles on an inch scale. Then measure off this inch on the fabric and remove a few threads each side of the points outside the given inch. The inch can then be separated and the threads counted individually.

Determination of Weight Per Unit Area

The weight per unit area is usually expressed in ounces per square yard or ounce per lineal yard. A square yard means of course 36" by 36" whereas a lineal yard is 36" by the width of the cloth. The weight per square yard can be dertermined from any sample of cloth, but if weight per lineal yard is to be calculated, the full width of the cloth must be known or a sample of full width suplied.

To find the weight per square yard, spread the cloth on a smooth surface taking due care to ensure that the cloth is free from folds or wrinkles. A template of exact known area (one tenth of a yard is a useful size) is then placed on the cloth and a piece of cloth corresponding to the area of the template is cut out, care being taken to ensure that one side is parallel to the selvedge. All that is needed is to weigh the cloth and multiply by ten, after having first dried the sample so that results will not be affected by moisture content.

A similar procedure is adopted to find the weight per lineal yard, except that the sample is cut to the full width of the cloth. The piece cut from the sample should be three, six, nine, twleve, eighteen or thrity-six inches in length so as to allow for easy calculation.

Distortion of Threads by Weaving

The student should note that a piece of cloth is always shorter than are the individual threads which go to make the cloth. In the process of weaving the threads are curved as they combine to make cloth due to the interlacement of the threads. This is not shrinkage as of course, if they are unravelled the threads are still the same length as before weaving, i.e., longer than the cloth.

Shrinkage

The first wetting of a fabric after weaving results in a relaxation of the tension which it has undergone in the process of weaving, and this results in shrinkage. Finishing processes also account for shrinkage. Shrinkage is always expressed as a percentage. To determine the shrinkage, a square, preferably twenty inches is marked warp-way and weft-way. The sample is then placed flat in a tray of water for two hours. The piece is then gently withdrawn from the tray and placed between two thick glass plates to dry. When it is dry, the marked square is again measured, and the percentage of shrinkage claculated.

Strength Testing of Fabrics

The strength of the fabric is dependent on the strength of the individual threads, concentration of threads, type of weave and the damage that the yarns or fabrics may have suffered during manufacture. Fabric strength can be assessed in several ways– the load required to break a strip of known width under controlled conditions, bursting tests or the pressure required to burst a fabric, wearing tests, usually of a rubbing nature, and chemical tests to find out the amount of chemical degradation. It should be stated that a yarn or fabric may be weakened either by mechanical damage (which will not affect its chemical properties), or by chemical action (e.g. the loss of strength due to exposure). In either case, breaking, bursting or wearing test will all give useful information.

The strength of a fabric may be tested by bursting, breaking or wearing.

Bursting Tests

Such tests are commonly used to determine the strength of knitted fabrics. A standard area of the cloth is subjeted to a gradually increasing pressure until the fabric bursts.

There are various types of machines, and the pressure may be applied by means of a piston, hydraulically or by compressed air forced into a rubber diaphragm until it bursts the cloth which is stretched over it.

Breaking Tests

Breakage tests are of great importance as a means of determining the strength of fabrics, and in fact are carried out as a matter of routine. The test commonly used in India is the strip test.

Strips of uniform size (usually $6\frac{5}{8}$ x $6\frac{5}{8}$) must first be prepared. Variations in size or humidity may lead to inaccurate results. The strips should be cut slightly oversize, and then sufficient threads removed from each side to give the exact size. Himidity during testing should be measured and a record kept. The strips should be accurately placed between the jaws and clamped by both leavers uniformly.

The machine most extensively used for this test is the Good brand Horizontal Machine. This registers on a dial the acutal breaking strength of the strip. The strip is placed between clamps on a flat plate. Power is applied and the clamps are steadily drawn apart against the pull of a pivoted, weight. The weight equivalent of the power needed to stretch the fabric is registered by a pointer on a dial. When the fabric finally breaks the needle stops and does not return to zero. The reading of the needle is the breaking point of the fabric in pounds. The stretch is registered in inches on a scale.

The machine may be worked by hand or by a small electric motor. Power drive is preferable, as it provides more even motion. The speed should be constant.

There are other machines available for tests of this nature, in particular the Schopper Vertical Strength Testing Machine which incorporates a graphic recorder to provide a permanent record. There are also combined yarn and cloth testing machines.

Wearing Tests

While the accepted standard of strength is tensile strength, it should be noted that of two cloths, the one with the lower tensile strength may have greater wearing qualities. Hence, tests of the wearing qualities of cloth are sometimes carried out. In these machine the cloth is subjected to friction from revolving knives, rubbing over abrasive blocks and being rubbed against itself.

10

TESTING OF KNITTED FABRICS

There are two types of construction in knitted fabrics:

(1) Weft knit, and

(2) Warp knit.

Weft Knit

This is the construction of a fabric in which a single yarn travels horizontally making a succession of loops. Knitting by hand is weft knitting. Both circular and flat machines are used for weft knitting. It is possible by means of the circular machine to form a tube with "wales" or the number of vertical loops the same throughout.

Warp knit

This is the construction of a fabric in which loops are made by numerous yarns simulating the warp in a loom. They form vertical rows of loops and there are as many yarns as there are needles used. The flat machine is generally used for this type of knitting. The one-bar or one-warp tricot, the two-bar or two-warp tricot and the milanese type are used in warp knitting.

Machine Needles

In a knitting machine, the number or needles in use for a fabric is equivalent to the number of vertical rows of loops of the fabric. Two kinds of needles are used in this industry-the latch needle and the spring-beard needle. The working of the latch needle is faster than that of the spring needle.

Thread

In silk hosiery, the term "thread" is used to describe the number of strands of raw silk. The higher the thread number, the heavier the yarn. Rayon and Nylon stockings are classified in terms of denier and not thread. The denier of rayon or nylon yarn is the weight in grams of 9,846 yd. or 9,000 metres. For instance, if 9,846 yd. of rayon weigh 40 grams, it is 40-denier rayon. If it is 90 grams, it is 90 denier rayon. So it is seen that the higher the denier numer the coarser the yarn.

Twist

Generally five turns to the inch are given to the yarn for a two or more thread silk stocking unless the term "high twist" is used. For example, a three-thread Chieffon means that three raw silk strands have been twisted to make a yarn having five turns to the inch. The resultant yarn is called a three-thread Tram. But if this yarn is again given another twenty turns per inch making twenty five turns per inch in all, it would become a "high twist three-thread tram"

Gauge

The term "gauge" means the number of needles in one and half inches of the needle bar used in the flat machine. A 45-gauge fabric means 45 loops or stitches in one and half inches. When counted horizontally this will mean thrity stitches per inch. These vertical rows of the fabric are called "wales". When the guage is higher, the loops are closer and finer.

Courses

The rows of handknitted stockings are built up horizontally marking a course. Similarly in machine knitting, the successive rows of loops are called "courses". They are counted vertically–so many "courses" per inch. Counting courses per inch is the most convenient method of measuring the quality of a knitted fabric. A good quality fabric should be neither too boardy nor too slack.

Yarn Used in Fabric

From all weft knitted fabrics sufficient lengths of yarn can be unravelled from the end for testing for count, twist etc. But it is very difficult to withdraw any reasonable lengths of yarn from a warp-knitted farbic. Accordingly great care should be taken to measure and weigh the short length available in order to find the counts by calculation.

11

THE TESTING AND ANALYSIS OF TEXTILES

Finishing of Textiles

Sizing and finishing are two distinct stages in the manufature of textiles. Sizing takes place before the weaving operation and entails the addition of binding materials such as flours, starches etc. to the thread. Finishing is the process which the woven fabric undergoes before sale, to give it a desired feel, shine, freshness, body and general "shop look" There are many types of finishes, including physical, chemical and mechanical. The modern trend is towards the production of durable and lasting finishes.

Finishing

Objects: The object of finishing is to add attractiveness or desirability to fabrics. There are also specialised finishes whose function is to make the fabric especially suitable for a particular purpose. Examples are waterproofing, crease resistant finished etc. Finishing is a branch of textile technology which has made great strides recently, and new finishes are coming into existence almost daily. Some of the better known finishes are dealt with below, but the student should familiarise himself with new developments as they arise by a careful study of trade periodicals.

TYPICAL FINISHES

(1) Beetling: This is a process whereby a row of wooden hammers fall on the surface of the material. This operation closes the fabric and imparts a soft glossy finish. Beetling leaves a characteristic "water-mark" on the surface of the cloth. This process is now obsolete. Even earlier, it was used mainly for Linen fabrics.

(2) Calendering: This process corresponds to domestic ironing. The fabric is passed between a number of rollers which may be steam heated. Calendering has the effect of smoothening the cloth and giving it a bright appearance. The main variations are Swissing., Frictioning and Chasing. The term Swissing refers to the normal calendering process. Frictioning produces a very high glaze obtained by imparting a higher surface speed to a heated metallic roll, well above the cloth speed. This frictional effort produces the very high glaze. Chasing involves running the fabric between soft nips mainly between two cotton bowls. A rather mellow glass is produced.

(3) Schreiner finish: This is a specialised form of calendering wherby the cloth passes between heated rollers which are engraved with fine paralled lines. The result is a lustrous finish due to the reflection of light from the small ridges formed in the cloth.

(4) Mercerised finish: This is a very important finish. The cloth is passed under tension through cold caustic soda solution, resulting in improved lustre.

(5) Filled finish: Cloth may be finished by the application of startch and filling materials to add stiffness and weight. Typcial stiffening substances apart from starch are glue, dextrin, case in. Filling materials which add weight to the cloth are Chinaclay, Talc and similar mineral substances. Back filling finish—a variation in which the fillilng mixture is applied to only one side viz., The back-side.

(6) Sanforising: This is a process of compressive shrinking. The potential shrinkage of the fabric is determined by conducting a shrinkage test with a sample. The

fabric is then compressed by the sanforising machine to the required length and width. The result is a practically un shreenk farbic. Other similar finishes are "Rigmel", "Evaset" etc. Only length shrinkage is positive. Width is stabilised only up to the relaxation level.

(7) Trubenising: The object of trubenising is to give permanent stiffness to a fabric, especially for collars, shirt cuffs, etc. The process consists of weaving a cotton fabric with alternate threads of cellulose in either the warp or the weft. The fabric is then moistened with a mixture of alcohol and acetone; on which heat and pressure are applied. The acetate rayon threads soften to a jelly and spread out to fill all the spaces between the cotton threads. The result when cool is a stiff fabric which is unaffected by washing.

OTHER FINISHES

Anticrease, Durable press

The purpose of these finishes is to impart crease resistance to rayon, cotton, or cotton-rayon fabrics. It consists of impregnation with DMEU (Dimethanol Ethylene Urea), or other related compounds or other using followed by heat treatment to cure or set the resinic compounds. A variant of this is the well known Durable Press and Permanent Pleat etc. in which the impregnated fabric is only dried and left incurred, this being carried out only after the garrments have been made. These are maintained in the shape pleats etc., required during the curing treatment. The set shapes pleats etc. are more or less permanently retained during use. These finishes are also sometimes known as "Min-iron" or "Non-iron" as these reduce the need for ironing verymuch.

Raising/Napping

The object is to raise a pile or nap or fibbers for soft effects on cotton or cotton-wool fabric. It consists of passing the fabrics against the surface of rotating teazle covered or wire covered rollers which teaze or pluck out fibres from the fabric surface to form a nap. In case of cotton, the weft is made with a very low twist to facilitate the raising. The process is often used to produce cotton flannelettes to imitate woollen fabrics.

Heat-setting

This is a special heat treatment which utilises the thermoplastic property of synthetic fabrics. The purpose is to eliminate the distorting and shrinking tendency of such fabric during processing and use. The usual temperature of setting is around 200 degree centigrade for filament fabric and lower temperatures for blends. Heat setting of Acrylic and blends containing Acrylics is avoided due to yellowey tendency and risk of damage. The fabrics must be maintained at the required dimensions for end-use, during the heat treatment. The set dimensions are permanent during use up to the temperature of setting.

The student should familiarise himself with the more common finishes and learn them when he seas them.

12

VARIETIES OF FABRICS

Textile fabrics are manufactured for innumerable purposes, according to the requirement and demand of the consumers. Some are used for garment material, some are for household use as sheets, towels, curtains and bedspreads. Others have industrial uses as filter cloths, hose, belting fabrics, tents, awnings etc. They are made in different widths. Generally cotton dress materials are thrity six inches wide. Draperies, and all other fabrics have their own width depending upon the subsequent use of the cloth.

Every cloth examiner should be familiar with the various types of fabrics that are manufactured. Listed below are the common materials with their main characteristics. Carpets, rugs, etc., though in sense are textiles, have been omitted as being a subject for separate study.

COTTON FABRICS

Drapery Fabric: This is generally 50 inches wide, used for covering armchairs, setters etc. It is made by dobby or Jacquard loom in order to produce the figured design.

Bath Robe Cloth: This is a thick, double-faced cloth, napped on both sides made by Jacquard. As its name implies it is used for bathrobes etc.

Boradcloth: This is made of combed or carded yarn. It is mercerised when made from combed yarn. It has a close plain weave having about twice as many ends as picks, producing a fine crosswise ribbed effect. This cloth is used for shirts, pyjamas, etc.

Buckram: This is made by glueing two fabric together making a stiff material. It is used for interlinings, and in millinery for stiffening hats.

Bunting Cloth: This resembles wool bunting. It is a plain weave fabric used for flags and similar decorations.

Byrd Cloth: This resembles raincoating. The cloth has a wind resistant -weave and is treated to make it water-repellent.

Calico: This is a plain woven printed fabric, slightly glazed.

Cambric: Cambric has a closely woven plain weave with a soft finish, slightly glazed. It is generally used for shirts, underwear, handkerchiefs, etc.

Canvas: It is a heavy fabric in plain weave. Sometimes it has coloured stripes. It is used for sails, awnings, etc.

Gauze Cloth: This is very light in weight, and has a loose construction. The weave is plain and the number of threads per inch is very low. It is used for bandages when bleached. It is also used for other purposes such as book binding etc.

Corduroy: The foundation of the cloth has a plain weave, but it has ridges running lengthwise. It is water-repellent and may be white or in colours. It is used for dresses, suits, upholstery,

Crepe: This is a fabric of plain weave which has a crinkled effect. Which can be produced by alternate mercerised and unmercerised stripes. It may be white, dyed or printed. It is used for dresses and lingerie. The cranked effect can also be produced by the use of very highly twisted or even crimped yarns.

Crinoline: This is highly sized with starch and glue. The weave is plain and the number of threads per inch is low. It is used for interlinings and in millinery.

Damask: This is made on the Jacquard loom. It has a figured weave which is reversible and a smooth finish.

Sometimes it is given a permanent linen-like effect and mercerised or schreinerised for table use.

Drill: Drill is strong fabric of coarse yarns. The weave is broken twill or left hand twill. It is white or coloured and used for suitings, uniforms, etc.

Denim: A strong fabric of twill weave preferable 2/1 made of 10s navy or blue dyed cotton yarn and a natural white cotton weft of around 16s counts. The weight is around 8 to 9 pz. per sq. yd. It is a hard wearing material and has become popular all around the world.

Duck: This is a heavy and coarse cloth generally made on weight basis such as 8-ounce or 16-ounce per unit area. Single or two-fold yarns are used. The various types of duck manufactured include army duck for sails etc., and single-pick flat duck for coated fabrics and shoe linings, wide duck for mailbags; and belting duck for conveyor belts.

Flannelette: This is a heavy, soft and napped fabric. It is used for infants wear, sleeping garments, etc.

Garbardine: This is a firm and closely woven fabric in right-hand twill weave, having approximately twice as many ends as picks. It is often mercerised and dyed and used for children's cloths, suits, etc., and water-repellent rainy day clothing.

Honey-Comb: The cloth has a rough surface in honey-comb pattern and is made from soft yarns. It is used for towelling.

Huckaback: This is made on a dobby loom with a small geometric weave. It is used for towelling, etc.

Jean: This is finer and lighter than drill. The weave is twill or herring-bone.

Longcloth: The textual of this cloth is closer than cambric. The weave is plain and close with or without sizing. There are various types of this cloth. It is used for shirtings, underwear, etc. Generally means, "Longcloth" bleached sheeting.

Mull: Mull is a fine plain weave cloth with a soft finish. Generally it is mercerised. It is used for ladies dresses, etc.

Organdie: The chief characteristic of this cloth is its crispness. It is a fine cloth in plain weave usually made from combed yarn. Permanent crispness may be given by chemical treatment; otherwise the crispness may be washed out. It is used for ladies dresses, etc.

Pique: The cloth has a special weave where warp yarns are used in groups producing a wale effect lengthwise. Extra backing threads may be used, which when caught by weft threads make the webs prominent. It is used for dresses, etc.

Poplin: Generally mercerized, this is closely related to broad cloth having a lower number of threads but with a pronounced weft rib. It is used for shirtings, etc.

Rep: This is similar to poplin but the cloth is heavier and has more pronounced weft ribs. The cloth is termed "armure" when it has Jacquard figure on a rep background. It is used for draperies and upholstery.

Sateen: This is a lustrous cloth in five-shaft twill weave generally weft-way but it may be warp-way also. It is fairly heavy and mercerised or schreinerised. It may be dyed or printed and is used for garments, etc.

Sheeting: This is similar to longcloth.

Suedcloth: The cloth is napped and has a double weave.

Terycloth (Turkish): This is a fabric with uncut loops on both sides. It may be white dyed or printed. It is used for towels and bath robes.

Ticking: This is a heavy twilled fabric with blue and white warp stripes. It is used for mattress and pillow covers, etc.

Velveteen: This is a fabric with pile weave where weft floats are out. It is used for dresses, decorative purposes, etc. Velveteen and corduroy are both weft-pile fabrics.

Voile: Voile is made from hard twisted yarns which produce a number of warp and weft threads. It may be coloured or printed—And used for dresses, etc.

Besides the above-mentioned fabric, there are others such as mosquito curtain cloth-having square or round meshes, cloth with leno weave etc.

WOOL FABRICS

Bedford Cord: This is a vertically ribbed worsted fabric used for coats, slacks, etc.

Broad Cloth: This is a soft fabric made form fine woollen yarns having a twill weave. It has a napped and polished surface. It is used primarily for men's and women's suits.

Bunting: This is a plain woven worsted farbic used for flags.

Cashmere: Real cashmere is made from cashmere goat hair but the name is also given to a lightweight soft woollen fabric of twill weave which is used for infant's garments.

Crepe: This is a lightweight worsted fabric in plain weave, having a crinkly surface. It is used for women's wear.

Flannel: Flannel is a soft fabric in plain or twill weave-generally 2/2 twill. It has somewhat napped surface to obscure the weave. Flannel fabric may be all cotton, all wool or mixed, but if the material is not specified it means, it is wool, Viyella flannel has a standard construction. It is a blending of cotton and wool in about equal amounts in both warp and weft. The shrinkage is controlled by the cotton. It is used for blouses, shirtings, and sportswear.

Gabardine: This is a firm worsted fabric in twill weave with a hard finish. The characteristics of this fabric are its fine steep diagonal lines. Like flannel, it also may be of cotton or rayon but unless defined, it is worsted.

Mohair: This is a plain weave fabric with cotton warp and mohair weft. Sometimes wool is also used in the warp. It both may be light or heavy. It has various uses.

Palm Beach Cloth: This is just like a mohair fabric with cotton warp and cotton and mohair weft. It is used for men's washable summer suits.

Serge: Serge is a worsted fabric of twill-weave with a hard finish and it is clear on both sides. The diagonal lines run from upper right to lower left on the right side of the fabric. It is used for dresses and suits.

Shepherds Check: This fabric may be woollen or worsted or a mixture with cotton or rayon. This is a very popular material for dresses and suits. It is a small check designed farbic usually in black and white, and the weave is twill.

Suede: This may be of various fibres unless defined as wool suede. It resembles suede leather due to its finish. It is a fine soft woollen fabric with clipped nap.

Tropical Worsted: This is a firm plain woven light-weight worsted fabric used for men's summer suitings.

Tweed: This term is applied to a variety of suiting fabrics of heavier type. Tweed may be either woollen or worsted and of different weaves. Woollen yarns are generally used for colour effects. Harris Tweed is an English product. The British Government Board of Trade defines. Harris tweed as follows: "Harris Tweed" means a tweed made from pure virgin wool produced in Scotland, spun, dyed, and finished in the Quter Hebrides and hand-woven by the Islanders at their homes in the islands of Lewish, Harris, Uist, Harra, and their several appurtenances and all known as the Quter Hebrides.

Unfinished Worsted: These fabrics are made of worsted yarns and are of twill weave. They are not sheared and so have a slightly napped surface where the weave is not seen clearly. They are used for suits.

Voil: This is light worsted fabric in plain weave having a firm clear finish. The yarns used in this fabric are lightly twisted. It is used for dresses.

Whip Cord: This is a fabric of worsted yarn in twill weave having a very pronounced diagonal cord. The fabric is closely woven and has a clear finish. It is used for riding garments, suits, etc.

RAYON AND SILK FABRICS

Broad Cloth: Sometimes this is called "Tub silk". This is a silk shirting in plain weave used for dresses also.

Broade: This is a rich fabric having surface patterns where extra weft yarns are used against a plain background. Metal threads are also frequently used in this fabric. It is used for elaborate dresses and for decorative purposes.

Crepe: Tightly twisted weft yarns are used in this fabric with S and Z twist in alternation.

Georgette Crepe: In this fabric crepe yarns are used in both warp and weft. It is used in dresses and lingerie.

Damask: Damask may be light or heavy. It has rich weave with small or large reversible patterns generally woven on a Jacquard loom. Light damasks are used for dresses and lingerie whereas heavy ones are used for draperies and upholsteries.

Moire: The term implies "Watered". This is a finish given to a corded fabric by pressure with heated engraved rollers resulting in a watered effect. The finish is not permanent to steam or water. Sometimes a satin back ground is used in moire fabrics. It is used in dresses. Suits and trimmings.

Satin: The fabric may be of all silk or backed with cotton or rayon. It may have a five, eight or higher shaft weave. It is used for dresses draperies, upholsteries, etc.

Taffeta: This is a fabric in plain weave. Firmly twisted organzine yarns are used in the warp to make a closely woven firm material. It may also be weighted. Sometimes warp and weft of different colours are used to produce an irridescent effect. It is very popular for dresses.

Velvet: Velvet is a pile fabric having a short pile. When the pile is longer than one seventh of an inch., the fabric is called plush. The fabric may be all silk or silk backed with rayon pile or cotton backed with silk pile. The background of the fabric may be plain, twill or satin. When the fabrics made with a silk back and rayon pile, a brocaded pattern can be produced by removing the rayon pile chemically. The fabrics used for apparel and decorative purposes.

Other Fabrics

Various types of fabrics are manufactured in linen. Example are cambric, canvas, damask handkerchief linen, and towelling. Sail cloth is manufactured from hemp. As hemp is not weakened by salt water, was almost irreplaceable for material that are in contact with sea-water. Now synthetics especially Nylons and Terylene (Dacron) etc. are being increasingly used for this purpose. Even sackings of jute are being replaced by polypropylene fabrics.

Asbestos is used widely as a fireproofing and insulating material. This fabrics is used for fire-proof curtains, in suits and gloves for firemen and industrial workers. Asbestos is a mineral product–a silicate of aluminium and magnesium which is capable of being spun and woven into cloth. Rayon has innumerable uses, even outside of the clothing field. Sharkskin is a very popular fabric made of acetate Rayon. It is heavy and firm and is generally of basket weave. There are also other Rayon fabrics such as Gabardine, Damask, Net, etc.

13

THE EXAMINATION OF CLOTH

The examining of cloth is of great importance both to the manufacturer and the wholesaler. The manufacturer must have a good system where each piece of cloth is inspected by a skilled "cloth looker" before despatch of the goods. This ensures that defective cloth is not placed on the market and that the manufacturer establishes a good reputation in the trade. Every cloth examiner must have a good knowledge of the price and the quality of the material he examines. He should examine the cloth with common sense. An ordinary bleached sheeting should not be judged by the same standard as poplin. A defect in one particular material may not be a defect in another type of material. Weavers and looms are not yet perfect and so faulty material is bound to come and at times cannot be avoided but there are some faults which are entirely due to carelessness on the part of the operators and these should be brought to the absolute minimum. There is yet to be a recognised standard for defects in a piece of cloth, but from 5 to 10 % material with minor defects should be allowed to go forward without any complaint.

Common Faults In Cloth:

The most common faults found n woven fabrics are as follows:

1. Bad selvedges.

2. Broken ends.
3. Read marks.
4. Broken picks.
5. Broken Patterns in Dobbies and Jacqnards.
6. Broken Patterns in checks stripes.
7. Thick and thin places in weft.
8. Colours in stripes and checks being the wrong shades.
9. Oil stains.
10. Rust and dirt stains.
11. Mixed weft.
12. Temple marks.
13. Shabby yarn.
14. Floats of yarn.
15. Neps in yarn or impurities.
16. Uneven yarn.
17. Mildew stains.

Bad Selvedges

The selvedge is very important and a bad one will always give trouble in finishing. Moreover a good selvedge is usually accepted as indicating a good cloth. The manufacturers should always try to avoid badly woven, broken in places, raggy and loose selvedges in the fabrics. Bad selvedges are always due to careless weaving.

Broken Ends

This is also due to carelessness on the part of the weaver who does not piece up the warp threads when they break, thus leaving a thin place in the fabric which is seen as a line. It may also be caused by careless warping or slashing.

Reed Marks

This is a manufacturing defect where the ends are grouped together in twos, threes, or fours, giving a very bare appearance to the fabric. Generally, a reedy cloth is harsher to the feel than one which is well covered. The defect should be avoided.

Broken Picks

This is another weaving defect which should be avoided by the weaver. This is caused by the breaking of weft tread while the loom is running. The defect is seen as a thin line where the weft is missing.

Broken Patterns

This defect appears when an end breaks in a fabric and is "tied up" to a wrong thread. It is seen as a line down their piece or two threads running together instead of one. The defect may be also due to improper working of a dobby or Jacquard.

Thick Places

This is another result or careless weaving. Here there are too many picks together caused by the reed, beating too hard against the feel of the cloth and thus forcing more picks than required. This is a bad fault and will appear more prominently after finishing.

Thin Places

This is also the result of careless weaving generally seen in the weft. The defect occurs when the weaver restarts the loom without seeing whether the new pick will flush properly with the last one at the time of weft breaking or running out. It may also be due to wrong starting of a loom.

Wrong Shade in Coloured Yarn

At times this may be a serious fault. Before using the coloured yarn, care should be taken to ensure that yarn of correct shade is being used.

Oil Stains

These stains cause great trouble in woven fabrics as they may come from spinning, weaving or finishing. Sometimes it is difficult to trace where the oil came from, particularly in a finished cloth. In case single thread are stained either in ward or weft, there the fault lies with the spinner but if patches are found in grey pieces then it is the weavers fault.

Iron Stains

This is a common stain in grey gods. When the yarn passes through the rusty reed of the loom it gets stained. Sometimes these stains are caused by water dropping from a rusty pipe. In finishing also, careless piling of damp cloth coming in contact it nails etc., causes these stains.

Mixed Weft

This defect is due to wrong use of weft by the weaver. It appears as a dark patch all across the piece which is intensified after bleaching or dyeing.

Temple Marks

These defects are seen in the form of small holes at either side of a piece caused by the pins on the temples. When the temples do not work smoothly as the cloth is being drawn round the cloth roller, there is a bragging action and the temple pins tear the yarn.

Uneven Warp or Weft Yarn

These defects occur in spinning. Carelessly spun yarn shows thick and thin places in parts. The very thick places are termed "Club" show up very badly in the finished fabric.

Floats

These are the result of ends breaking and becoming entangled with others and leaving a loose piece of warp or weft on top of the piece. They are a bad fault.

Neps and other Foreign Matter in Yarn:

Neps are the short, unripe fibres, which are not affected by bleaching or dyeing. In a dyed cloth, dead and unripe fibres are seen as white speaks.

Broken leaf and seed in yarn also give trouble in bleaching and dyeing and they are particularly visible in light shades.

Storing Damages

Heavily weighted silks become tender when stored for some time, due to the action of the mineral compounds used in the finishing. White cotton goods have a tendency to turn yellow. Coloured cloth is altered by atmospheric changes. When cotton cloth is stored in a place where there is a lack of air, there is a tendency for the growth of mildew particularly in damp rooms.

Mildew

Grey goods are more liable to mildew than bleached or dyed goods. Bleaching or dyeing have a destructive action on the mildew growths. The microscope is very helpful in the determination of mildew in fabric. There are three types of mildews which are:

1. Green mildew.
2. Brown mildew.
3. Yellow mildew.

Green mildew is seen in rather large patches on the fabric. Brown mildew is often taken to be rust stains, which takes the shape of small circular spots. Yellow mildew is found as patches and spots and is very common in cotton cloth.

14

TESTING OF BLEACHED FABRICS

Introduction

The various processed entailed in bleaching and dyeing textile fibres constitute a distinct subject for specialized study. However, we give here brief notes on the more important processes in order to facilitate the study of the testing of bleached and dyed fabrics.

Scouring and Bleaching

Scouring is the process of half bleaching, which has as its object the removal of impurities such as oil and dirt in fibres. Scouring does not affect the residual colouring matter from the yarn or material, thus giving a good white appearance.

These processes, especially bleaching, depend for their success upon strong chemical action, and care must be exercised that the bleaching or scouring agent used stops at impurities and colouring matter and does not go on to alter the basic properties of the fibre thereby impairing its strength, elasticity or lustre.

Scouring is usually carried out by means of soaps, alkalis, petrol or benzene. Bleaching is carried out by means of a wide range of different substances, the factors governing the choice of a particular bleaching agent being the nature of the fibre, the degree of bleaching required, cost, etc., It should be noted that, a plentiful supply of suitable water is a necessity for both scouring and bleaching.

Bleaching Agents

There are two distinct types of bleaching agents. These are Reducing bleaching agents and Oxidizing bleaching agents. Reducing agents bleach the fabric by producing nascent hydrogen while oxidizing agents use nascent oxygen either directly or indirectly. It should be noted that few chemical change affected by oxidizing agents is more or less permanent, but in the case of reducing agents the articles bleached, mav yellow with age owing to the effect of atmospheric oxygen which reversed the original reaction.

The oxidizing bleaching agent most extensively used for the bleaching agents are chlorine, persulphates and ozone.

The oxidizing bleaching agent most extensively used for the bleaching of cotton is bleaching powder which liberated chlorine. The quantity of chlorine liberated indicates the quality of the bleaching power. A good sample should contain from 35% to 38% available chlorine. Bleaching powder is not suitable for bleaching animal fibres.

Linen is partly bleached by exposure to air. The exact causes of the bleaching are a matter of some conjecture but it is thought that air has little to do with it. The ultraviolet rays present in diffused sunlight are the principal contributor to the actual bleaching. It is however, thought that the traces of ozone and hydrogen peroxide present in the atmosphere have something to do with the process.

The Bleaching of Wool

The principal impurities present in raw wool are wool fat, dried perspiration suint dirt and vegetable matter which has become entangled with the wool. Most of these are removed by the initial scouring processes and the amount of bleaching which is subsequently necessary, is dependent upon the thoroughness of the scouring. Many types of wool require very little, if any bleaching before dyeing and are merely tinted or "dyed white". Where necessary, wool is bleached with sulphur dioxide, sulphurous acid or hydrogen peroxide.

The Bleaching of Cotton:

Cotton bleaching is a specialized craft, quite distinct from any subsequent dyeing operations. Briefly the processes are as follows. First the cotton may be steeped in water for some hours. This softens and dissolves impurities soluble in water. Next the goods are washed. Then follows lye boiling. This consists of boiling in a dilute alkali solution usually under pressure in a closed vessel called a kier. This process dissolves most of the impurities which are subsequently washed away. It should be noted that lye boiling gives rise to alterations in weight, length, count twist and tensile strength the last named being increased. Then follows the actual bleaching. The goods are passed through an aerated solution of bleaching powder, either the goods or the liquor being mechanically agitated. When bleaching is complete, the goods are soured by passing them through a dilute solution of hydrochloric or sulphuric acid. Finally they are soaped. Coloured goods require special treatment to prevent bleeding and marking-off of the colours.

Properties of Bleached Cotton:

Well bleached cotton should comprise almost pure cellulose. The tensile strengths should be at least the same as that of the original cotton. There should be no harsh feel. Bleached cotton should not contain acid, residual chlorine or unremoved non-cellulose impurities. The whiteness should be pure and durable.

Analysis of Bleached Goods:

Bleached goods should first be examined for colour, tensile strength, moisture content, presence of mineral impurities etc.:

(1) Colour: Colour defects in bleached cloth are either a yellowish or bluish shade. The simplest and most practical test is to compare with a standard white sample. Cloth which comes form the bleached process looking yellow can have the fact concealed by using blue. Most of the blue colours used are soluble in water, and hence if it is suspected that blue has been added, the sample should be washed in cold water and dried. Any yellowish appearance after washing will confirm that the sample has been blued.

(2) Tensile Strength: After allowing for the increase in strengths produced by lye boiling carefully, bleached cloth should show no reduction in strength. Tensile strength after bleaching would be compared with tests taken in the grey state to show whether there has been any tendering.

(3) Moisture Content: This can be assessed as previously explained.

(4) Mineral Content: Bleached cotton should consist of almost pure cellulose. The presence of excess calcium carbonate reveals that souring has not been properly carried out, A harsh feel in bleached cotton may indicate that there is excess calcium carbonate present. This can be easily confirmed by burning a sample in a porcelain basin until nothing is left but a little white or grey ash. A drop or two of strong nitric acid is added to assist complete combustion. The ash of well bleached cotton should not be more than 1% by weight of the original sample.

Bleaching Faults in Cotton Cloth:

It is difficult to trace the cause of damage or discouloration which takes place during the bleach processes as there are so many factors which may contribute to such faults. Among these are water, chemicals, mildew, size used on the yarn, etc.

The commonest defects which should be noted are as follows:

(1) Mechanical damage caused by faulty bleaching machinery: Regularity of discoloration or actual holes at regular intervals may point towards mechanical damage. These can rarely be eradicated and necessitate overhaul of the machinery. Yarns damaged mechanically have a distinctive appearance under the microscope when stained with Congo Red.

(2) Mineral oil stains: These generally come from the oil used to lubricate the machines. They may be removed with petrol or benzene, or by milling with soap. They can usually be avoided if machines are kept clean and are not over lubricated. Most machinery used in textile processes is also designed that oil should not be splashed on to cloth, provided that the makers', instructions are carefully followed.

(3) Lime Stains: These usually have their origin in the kiers during the lye boiling process and result from bad boiling or plaiting. By the use of caustic soda boiling, these have been practically eliminated.

(4) Iron Stains: The commonest cause of such stains, particularly where maintenance is insufficiently supervised is rubbing against rusty machine parts and iron pipes. They may also be caused by the presence of iron in the water or in the chemicals used for bleaching. Strong concentrated oxalic acid solution removed iron stains.

(5) Lead Stains: These come from rubbing against dirty lead pipes. They are not easy to remove but may dissolve in very dilute nitric acid or ammonium acetate.

(6) Soap Stains: Calcium and magnesium soaps which have neen left in the fabric will trun yellow or brown on exposure to air. This process takes time and the fault may not be immediately apparent.

(7) Natural Oil and Wax Stains: Residual oil and wax, the result of inefficient bleaching may cause yellowing after a time. This can be detected by steaming the fabric.

(8) Tenderness: There are various causes of which the following should be noted :

(a) Oxycellulose produced in the lye boil.

(b) Overbleaching.

(c) The presence of chloramines.

(d) Hydrocellulose due to the presence of unremoved mineral acid.

Tests to Estimate Tenderness in Fabrics:

There are three important tests which are carried out to test for the presence and/or estimate the extent of tendernes in fabrics. These can be used for any fabric which has cellulose as the chief ingredient. The tests are:

(a) The methylene blue test.

(b) Viscosity or fluidity measurements.

(c) Copper number determination.

The Methyline Blue Test

This is a test which reveals the presence of oxycellulose in a fabric. Oxycellulose has a greater affinity for methylene blue than cellulose. When a sample is treated with methylene blue in a slightly acid bath, the tendered portion-i.e., that containing oxycellulose-absorbs more of the methylene blue than the untendered prortion, and hence, shows a darker blue.

Measurement of Fluidity

The basic principle governing this test is the fact that the fluidity of a solution of cotton dissolved in cuprammonium varies in accordance with the degreee of degradation of the cotton sample dissolved. In other words, if the given sample has been damaged through faulty bleaching or any other cause, so that the cellulose which goes to make up the cotton is changed in part to oxycellulose or hydrocellulose, the sample will show a greater fluidity. If the original sample was undamaged, the fluidity is less. The fluidity of a solution of the bleached cotton is tested in a viscometer and the readings compared with those given by a solution of a sample of the same cotton in the grey state. The greater the fluidity the more the tendering. In normally bleached cotton the fluidity should not be above 5 poises measured under standard conditions.

Poises =1/V, V = is the viscosity measured in dynes per sq.cm. per cm. per second.

Copper Number

Alkaline solutions containing copper are capable of being reduced by degraded cellulose. When tendered cotton is heated with Fehlings solutions, a red precipitate of cuprous oxide is formed, Undamaged cellulose does not give this precipitate. The degree of tendering can be assessed from the amount of precipitate formed.

15

TESTING OF DYED FABRICS

As in the foregoing section of the lesson, before dealing in detail with the various tests which should be carried, out, we shall first give brief definitions and descriptions of the various processes entailed.

Dyeing is the process whereby colour is imparted more or less permanently to textile fibres. Modern textile dyes are mainly organic compounds created in the chemists laboratory. The ancient dyes prepared by traditional methods are rarely in use.

The principal classes of dyes are as follows:

Direct colours for cotton
acid dyes
basic dyes
mordant dyes
sulphur dyes
vat dyes
insoluble azo colours or azoic colours
aniline black
reactive dyes, etc.

Direct Cotton Dyes

These are soluble in water and the cotton takes up the colour directly from their solutions and retains the colour fairly permanently. The fixation to the material of the dyestuff is physical rather than chemical. They include a wide range of colours.

Basic, Acid and Mordant Colours

These are mostly soluble in water. Cotton requires to be mordanted with either acid or basic mordants before it can be dyed with these colours. The basic colours are very bright and give deep shades with the use of a relatively small amount of the dyestuff. They are not very fast, and are destroyed by exposure to light over a period. Acid colours are bright but have little fastness on cotton. Hence, these two colours are used principally in dyeing wool and silk, where they form chemical combinations with the fibres thus giving fast shades. Mordant colours give relatively fast shades even on cotton. The cotton is treated with mordant solution and the colour is finally fixed on to the fibre by precipitation. The full process is rather complicated, and hence these colours are being replaced by vats, azoics and similar dyes.

Sulphur Dyes:

These are used extensively for cotton piece goods. They are insoluble in water but are dissolved on reduction with sodium sulphide. After dissolving, the dyeing is carried out by a process similar to that used for direct colours. The dyed material is subsequently oxidised on exposure to air, fixing the colour firmly in the fabric. The shades are dull and hence, these dyes are sometimes topped (over-dyed) with direct colours. These colours do not include red.

Vat Dyes:

Vat dyes utilise a similar principle to sulphur dyes, though they do not contain free sulphur. They are very fast to light, washing and bleaching and are used extensively for dyeing cotton, linen and rayon. Their cost is rather high and hence, they are used mainly for good quality goods.

Insoluble Azo Colours or Azoic Colours:

These depend for their action upon a fairly complex

chemical reaction in which two components are coupled chemically in the fabric to give a bright shade which is reasonably fast to washing, soaping and in most cases bleaching. The reds in this class are particularly bright and the range contains all shades except sky blue and bright greens.

Aniline Black

Though only one colour, this is distinct chemically from the previously mentioned colours. It is formed from oxidised aniline salt which gives a deep black shade which is cheap, bright and fast.

In addition there are the mineral colours such as prussian blue, chrome orange, mineral khaki. These are little used since the introduction of coaltar colours with the exception of mineral khaki. This gives cheap fast shade suitable for military uniforms and stores. As the process entails the deposit of mineral substances on the material it gives a harsh feel to the cloth.

Reactive dyes: This is a more recent class of colours. They combine chemically with cellulose and give bright shades fast to light and washing but not to bleaching.

Colour Notation

The three attributes of colour are hue, value and chroma or intensity. Hue is the name of the colour. Value indicates the lightness or darkness of the hue. In this connection the difference between tints and shades should be noted. A tint is produced by adding white to a hue, while a shade is produced by adding black to a hue. Chroma or intensity indicated the strength or weakness of a hue. For example, a sample dyed in a weak red dyebath has a weak red chroma, while on the other hand a sample dyed in a strong bath has a medium or strong chroma.

TESTING OF DYED MATERIAL

The principal tests which are carried out on a dyed fabric are for fastness to light, to washing, to bleaching, to perspiration, to boiling, to ironing and in the case of cloth for waterproofing purposes, to water. When carrying out such tests, it is important to bear in mind the purpose for which the material has been manufactured. Tests carried out on good quality material which is sold at a relatively high price, will obviously be more stringent than those carried out on cheap cloth.

There are certain standard tests used to identify the dye used in a fabric. We give below the tests for a number of common dyes.

Indigo Blue: If a few threads of the dyed material are boiled in a solution of hydro-sulphite, the colour will change from blue to yellow. The original colour will return when the threads are exposed to the atmosphere. If a small sample is placed in chloroform, the solution will turn blue if indigo is present.

Alizarine Red: Material dyed with alizarine red turns yellow if spotted with nitric or hydrochloric acid. If treated with a solution of caustic soda it if turns violet.

Turkey Red: This turns to violet, if treated with ammonia.

Benze Red: If sprinkled with nitric acid, the sample turns blue-black.

Congo Red: This turns violet or deep blue black on contact with nitric acid.

Common Yellow: This turns blue black on contact with nitric acid.

Longwood Black: Hydrochloric acid sprinkled on a sample dyed with longwood black turns it red.

Sulphur Black: When the sample is boiled in a solution of hypochlorites it turns brown, but regains its previous colour on exposure to air. (also see test below under sulphur colours).

Aniline Black: Turns dark green on contact with sulphurous acid.

Alizarine Black: Turns reddish in the presence of any acid.

Chrome Yellow: Turns black in any alkaline sulphide.

Chrome (Mineral) Khaki: Turns green on contact with concentrated hydrochloric acid.

Direct Colours

Boiling the sample with hydrosulphite solution removes the colour completely. The darker shades require boiling for some time before the colours is completely destroyed. Direct colours bleed freely on boiling in soap water except in the case of some of the lighter shades which however bleed on a second boil. After-treated colours are faster but are destroyed on bleaching.

Developed Colours

Good washing and fairly good light fastness. The colour is destroyed by hydrosulphite, though the process takes longer than in the case of direct colours. They do not bleed on second boiling.

Basic Colours

Fairly fast to washing but rather low fastness to light though this varies to some extent with different basic dyes. Hydrosulphite removes the colours though these return sightly on exposure to air. It should be noted that the return of colour is not so marked as in this case of vat and sulphur colours. A solution of bleaching powder in cold water removes the colour of basic dyes. Another test is to heat the sample slightly in methylated spirit. Basic dyes tint the spirit. It should be noted in this connection that some vat dyes also tint methylated sprit.

Sulphur Colours

These are fast to light but not to bleaching. They are little affected when boiled in water. They lose colour on boiling in hydrosulphite but regain the colour on exposure to air. All sulphur colour give off sulphuretted hydrogen when treated in a hot solution of stannous chloride. This turns lead acetate paper black.

Vat Dyes

These colours are fast to bleaching. They are unaffected by bleaching powder, but change colour when boiled in hydrosulphite, which is however restored on exposure to air.

Azoic Colours

Mostly Naphthol as series or its equivalents. These are stripped when boiled with hydrosulphite. The colour is not restored by exposure to air. They are fast to bleach.

Reactive Dyes

These are also stripped by boiling hydrosulphite solution and the colour is not restored by air-exposure. However, they are not fast to bleach but have a much higher wash fastness than direct colours and hence do not bleed in hot soap solutions.

N.B.: When using hydrosulphite a few drops of caustic soda solution are added in all cases.

To Distinguish Piece Dye and Yarn Dye:

If any difference in shade is noted between the warp and weft threads when the cloth is unravelled then, the cloth is yarn dyed. On the other hand it should be borne in mind that yarn dyed material does not necessarily have any difference in shade between warp and weft threads. A selvedge coloured from the rest of the cloth may be taken as an indication of dyed yarn.

If on unravelling a lighter shade is observed where a warp thread crosses a weft thread, the fabric is piece dyed.

16

TESTING OF FASTNESS

Dyes are rarely equally fast to all influenced and quite often a dyestuff of good fastness in one respect e.g., light is not found as good to other influence e.g., washing. This is quite common in the Solopheny Direct dyes series which are fast to light. Fastnesses are expressed in terms of numerical scales. For all properties except light the scale 1-5 is used. For light the scale is 1-8. The following gradations are in use. The dyed cloth is braided i.e., twisted together with white cloth for testing.

ALTERATION OF SHADE DURING PRESCRIBED TEST GRADES:

5. Shade unaltered.
4. Very slight loss in depth of alteration.
3. Appreciable loss or alteration.
2. Distinct loss or alteration.
1. Great loss or much alteration.

STAINING OF ADJACENT MATERIALS GRADES:

5. No staining of adjacent white.
4. Very light staining of adjacent white.
3. Appreciable staining of adjacent white.
2. Deep staining of adjacent white.
1. Adjacent white dyed deeply.

Fastnesses to Different Colour Fading Elements or Conditions Light Fastness Grades

8. Maximum fastness.

7. Excellent fastness.
6. Very good fastness.
5. Good fastness.
4. Fair fastness.
3. Moderate fastness.
2. Slight fastness.
1. Poor fastness

N.B.: Gradations based on standard practice.

Fastness to Perspiration:

Perspiration may be acid or alkali, and hence both acid and alkali tests should be made. The sample should be thoroughly wetted in an acid solution and then placed in a corked test tube. The tube is then kept at 35° C approximating to human body temperature for a period of twenty-four hours. The same process is then carried out after wetting in an alkali solution. The actual solutions for acid; test is 100 gms. common salt and 50 gms. glacial acetic acid. The alkali solution is made in the proportion of 100 gms. common salt and 10 gms. sodium carbonate, the solution in each case being made up to one litre by the addition of water. After treatment the sample should be compared with the original cloth.

The best test of all, however, is wearing as there are other factors present in human perspiration not fully accounted for by the above tests. It should be noted that almost all colours are affected to some degree by prolonged contact with human perspiration.

Fastness to Washing

To test for fastness for washing the sample is twisted together with an equal quantity of white cloth and soaked in a natural solution of soap and water at temperature of 50 degrees C for about twenty minutes. The sample is then squeezed out and dried. If the colour is fast to washing it should be unchanged after this test. Moreover, the colour should not have bled into the white cloth, which should be unmarked.

Fastness to Boiling

This is carried out in a similar manner to the washing test. The solution comprises 5 grm. Marseilles soap and 3 grms, soda ash per litre. The sample is boiled for 30 minutes in the solution and cooled in a bath at 40 degree C for a further 30 minutes. The result is then compared with the original.

Ironing Test

The dyed sample should be covered with two pieces of thin bleached and unfinished cotton clothe damped with boiling water. It should then be ironed out with iron hot enough to singe wool. The iron should be used until the covering cloth is dry. The ironed portion should be compared with original cloth and in addition, the white covering cloth should be inspected for any indication of bleeding.

Fastness to Bleaching

The coloured sample should be twisted together with pure white cloth and treated for one hour at about 65 °C in a bath of chloride of lime with 1 % free chlorine in a litre of water. It should then be rinsed, squeezed and dried and compared with the original.

Fastness to Light

This is a very important fastness required in colours. If a colour does not fade when exposed to northern light for one month in summer, it is said to be fast to light. This test is based upon conditions in the northern hemisphere, and must be modified elsewhere. In any case, the test is rarely carried out in actual sunlight conditions now a days, as a month is far too long to spend on one standard test, when substitutes for sunlight can complete the test in 48 hours. A mercury vapour lamp or violet carbon reproduces the same effect as sunlight. The sample is exposed to the direct rays of one of these lamps for 48 hours, and the sample is then compared with the original cloth.

Waterproofing Test:

To test for waterproofing the sample is fixed on a frame so that the material forms a hollow in the centre. Water is placed in this hollow and if the water-proofing has been properly carried out, there should be no drops on the underside after 24 hours.

When dyed cloth is braided (i.e., twisted together) with white cloth for the purpose of testing fastness, the following main standards of fastness should be noted.

(a) The colour of the sample is faded and the white cloth is coloured.

(b) The colour of the sample is faded and the white cloth is unaffected

(c) The colour of the sample remains unaffected and the white clothe is not coloured.

Mineral Khaki

There is a special series of tests for mineral khaki, which is used for military purpose and must be fast to laundering, perspiration and sunlight. Khaki material is used under far more arduous conditions than most other cloths. The tests are as follows:

Fastness to Washing :

The sample is boiled with a weak solution of sodium for 30 minutes. If it is unchanged at the end of this period it is fast to washing.

Fastness to Air and Light:

The sample is soaked for 30 minutes in a cold solution of hydrogen peroxide and should remain unchanged.

Fastness to Laundering and Perspiration :

The colour must remain unchanged after the sample has been soaked in lactic acid solution for 24 hours.

17

SPECIFICATION OF TEXTILE STORES

We give below some typical specifications for textile stores.

Specification No. 1— Cotton wool absorbent

(1) The coton wool shall confrom to the sealed sample, in all respects. It shall be tasteless odourless, and should have no acid or alkaline reaction. The ash content shall not be more than 5%.

(2) The cotton wool shall be of the cleanest type and fully bleached, and should be free from grease and other ingredients used in the bleaching process. It would also be free from dust neps, and other impurities. Felting and malting of the fibres are very objectionable and great care should be taken so that the fibres are not tendered or weakened in the process of bleaching.

(3) During the manufacture, no colouring matter should be employed to improve the appearance of the material.

(4) The cotton wool shall undergo a thorough carding process to produce a fleecy mass of fibres. The material shall be supplied in rolls of one pound packets. Suitable inter-leaved papers shall be used in the packing of the material.

(5) The power of absorbency of the cotton wool shall be through and immediate. It should sink in cold water almost immediately by its own weight.

Note: The student should note that though the name of the material is "Cotton wool absorbent". There is no wool present in any form in the material. In the industry, this name of Absorbent cotton is "Cotton wool Absorbent".

Specification No.2— Cotton sewing thread

(1) The cotton-sewing thread shall be manufacuted from the best quality of cotton. It shall be evenly twisted and be of even thickness throughout the length. It shall be free from knots, short lengths and other defects of yarn.

(2) In the case of coloured thread, the shade shall be unifrom and shall stand all the tests for fastness of dye satisfactorily.

(3) If it is khaki dyed, the thread must stand the boiling test. The test will be carried out by boiling the thread in a 2 % solution of soap and water for 2 hours.

(4) The thread shall be wound in reels of well seasoned wood. The size and shape of the reel shall conform to the sealed sample. Every reel shall be labelled with the name of the suppliers, the construction of the thread, the number of yards, of thread wound on the reel.

(5) The finish of the material shall be equal to that of the sealed sample.

(6) The particulars of the thread shall be as follows.

Quality No	No. of Strands	Counts of yarn	Breaking strength under air dry condition	No. of yd. reels
1	7	40s	125 oz	300 yd.
2.	5	30s	70 oz	300 yd.
3.	3	30s	60 oz	300 yd.
4.	3	25s	40 oz	300 yd.

(7) An average of 5 breaking tests for tensile strength shall be taken. A distnace of 20" between the grip shall be used in the test.

Note: The above is only a specimen of a specification for sewing thread cotton, to help the students.

Specification No.3—Cotton twine-5/10s

(1) The twine shall be manufactured from the yarn produced from good, clean, and long stapled cotton. The yarn composing the twine shall be evenly spun.

(2) The twine shall conform to the advance sample. It shall be well-twisted, and free from many knots and other defects.

(3) No sizing or dressing should be used in the manufacture of the material and it should be neatly and tightly would to produce well formed balls.

(4) The particulars of the twine shall be as follows:

Counts of the component yarn	No. of single thread	Twist for 2" of twine	Breaking strength twine	Type of delivery
10s	5	26	500 lb.	In ball of approx.30z

Specification No.4 — Dussuti-the Bleached-Cotton

(1) The fabric shall be manufactured from yarn which is evenly spun from good quality of cotton. The fabric shall be fully bleached to effect the maximum degree of whiteness possible for the type of cotton used in the manufacture, without causing tenderness in the fabric.

(2) The fabric shall be reasonably free from weaving faults, and other stains.

(3) An average of 5 test results shall be taken to determine the breaking strength of the fabric both warp and weft way. The test samples shall be 6 and 5/8" wide and the distance between the grips shall be 7".

(4) The particulars of the finished fabric shall be as follows:

(a) Weave-Plain

(b) Width-45"/45 and half"

(c) Minimum threads per inch
— Warp = 32 double ends
— Weft = 30 double picks.

(d) Weight per sq.yd. — 6 oz

(e) Breaking strength under normal air-dry condition
— Warp = 450 lb.
— Weft = 400 lb.

Specification No.5—Serge–the Blue–Worsted

(1) The fabric shall be produced from a quality of worsted yarn, which shall be evenly spun from wool, not inferior in quality than the sealed sample. The fibre shall be pure wool and any admixture of cotton, or other recovered or shoddy material is highly objectionable. The finish, thickness and shade of the fabric shall conform to the sealed sample and the fabric shall be reasonably free from streaks, stains and other weaving defects. It shall stand the test for fastness of dye, to exposure to air, light, washing, perspiration and ironing, and steaming.

(2) The particulars of the fabric shall be as follows:

(a) Width - 54".

(b) Threads per inch -

(1) Warp = 72 (2 fold worsted).

(2) Weft = 70 (2 fold worsted).

(c) Weight per square yarn for a 16 % regain - 11 oz.

(d) Breaking strength under normal air dry condition warp/425 lb. weft/325 lb.

(e) Size of test strips 6 and 5/8" x 6 and 5/8".

(3) Shrinkage of more than 2 % either in warp or in weft is objectionable. The shrinkage test shall be carried out by soaking the samples in cold water for two hours. The samples shall be lightly wrung by hands and then dried on a flat surface.

(4) The test strip shall be cut 7" wide both in warp and weft and then frayed on both sides, taking equal number of threads to reduce the sample to 6 and 5/8" width. Five tests shall be taken and then average made.

(5) To determine the weight per square yard, one sample of one foot square shall be cut. The sample then shall be conditioned to bone dry weight at 220° F. to 230°F. An addition of 16% of this weight shall be added to determine the weight per square yard for a 16% regain.

Specification No.6 — Gunny-the Jute-cloth

(1) The gunny cloth shall conform to the sealed

sample. It shall be manufactured from the same quality of the sealed sample. The textile finish and general appearance shall not be inferior to the sealed sample.

(2) The particulars of the supply shall be as follows:

(a) Weave - Plain.

(b) Construction - 10 porten (double ends) per inch.
12 shots - (single ends) per inch.

(c) Width - 20".

(d) Weight per sq. yard = 18 oz.

(e) Twist per inch of warp = 3.

(f) Twist per inch of weft = 2.

(g) Counts of yarn - warp = 5s approx.
weft = 4s approx.

Note: 300 yd. to a hank is the basis of the count of warp and weft, used in the manufacture of the gunny cloth.

(4) The material shall have a yellow strips, composed of two yellow ends, which should run along the length of the fabric. The stripe shall be one inch away from the selvedge.

Specification No.7 — Flax canvas

(1) The fabric shall be manufactured wholly from flax. The yarn shall be properly twisted. The flax used in the manufacture of the material shall be free from blacks. No weighting material shall be used in the manufacture of the canvas, which shall be uniformly woven. It shall be reasonably free from weaving flaws, and shall contain well-formed selvedges.

(2) The supply shall conform to the sealed sample and the particulars shall be as follows:

(a) Warp threads per inch = 25 double ends.

(b) Weft threads per inch = 22 single picks.

(c) Weight per square yd. = 26 oz.

(d) Breaking strength in saturated condition:
Warp = 425 lb.
Weft = 425 lb.

(3) The size of the test strips shall be one inch wide and the distance between the grips shall be 7".

18

THE TESTING AND ANALYSIS OF TEXTILES

Defects in Textile Goods

There are a number of defects which the student should be able to identify readily and know the cause.

Defects in Yarn

The first point to consider in assessing faults in yarn is the ultimate use to which the yarn is to be put. For example poor tensile strength might justify the rejection or a lot of yarn which is to be used for warp threads, whilst the same yarn might be eminently suitable for use as a weft. Faults in yarn may arise from a number of causes, not the least of which is the raw material. An unduly high proportion of unripe of immature fibre in raw cotton will result in weak yarn. Similarly, wool from diseased sheep may contain a high proportion of kemps. As explained earlier, microscopic examination will usually reveal such defects.

Yarn may be defective as a result or ill-adjusted spinning machinery. It the tension in the spinning bands is not uniform there will be thick and thin place. The number of turns per length of yarn must be uniform as any variation in this respect will result in weak places. And it should be remembered that the maximum tension which may be put on yarn during weaving must correspond to the tensile strength of its weakest point or breakages will occur.

If too much strain is placed on the yarn in the spinning process, the resulting yarn will be tender. Imperfect travellers and bad adjustment of ring spindles will result in poor yarn. When lower grade yarn is being spun a slightly higher twist is given to provide additional strength compensating for short staple and weak fibre. However, when the tension is removed the yarn tends to form loops.

Weft yarn for grey materials generally comes from the spinning department ready for the loom. Warp yarn must however go through the processes of winding, warping, sizing and drawing–in before going to the loom. These can give rise to weaknesses caused by excessive tension during processing. Bad sizing or overdrying may produce brittle yarn. On the other hand insufficient drying after sizing may encourage the growth of mildew resulting in tender yarn.

Defects in Woven Fabrics

This subject has already been dealt with in detail. However, an item which must not be overlooked in the inspection of woven fabric is the selvedge. Although the selvedge plays little part in the ultimate use of the fabric. It is of great importance when the cloth is to be finished. The primary object of the selvedge is to take some of the strain off the warp threads during processing. Dyeing, bleaching, printing and finishing all require strong selvedges. The selvedge should be strong, elastic and of a suitable weave. It should ensure that the strain is uniform throughout the body of the cloth. Unless is elastic the selvedge will split. It should not be hard and round but it should be flat and tape-like. Special selvedges required for voiled and venetians and also for drills which are to be dyed khaki. There should not be any tendency towards curling in as this can cause endless trouble and often damage in processing.

Defects in Bleached Fabrics

These defects have already been discussed earlier. In the finishing of high class fabrics, it is better to avoid high pressure in calendering to avoid risk of damage specially in Rayon goods.

Defects in Dyed Fabrics

In well dyed fabrics, the colour is uniformly distributed. A dyed fabric must not contain a higher or darker portion or patchiness. The dyeing operations should not affect the inherent properties of the fibre or fibres of which the fabrics may be made.

(a) Defects due to inherent faults of the fibres.
(b) Defects due to manufacturing operations of the fabrics.
(c) Defects due to faulty scouring or bleaching.
(d) Defects due to faulty dyeing.

It is generally the practice to hold the dyer responsible for all defects that are seen in dyed fabrics. Actually, he should be made responsible only for the defects that occur during the dyeing operations. It is not possible for the dyers to cover up the defects that are already present in the fabrics, received from dyeing. Rather in some cases, the dyeing makes the defects more prominent. So it is always advisable for the dyer to get the goods examined carefully to enable him to keep aside the faulty material for reference.

(a) Defects due to the faults of the fibre : The affinity for dyestuff and the degree of penetration is not the same in the different kinds of wool. Low grade wools possess considerable natural colour and they are not so easy to dye in light shades, or bright colours like high grade wools. So the presence of even a small proportion of low grade fibres may sometimes make it very difficult to get a perfectly even colour on the dyed fabrics. Spots or patches may be due to kemps or diseased fibres. Wool affected by weathering or bacterial action also possesses an altered affinity for dyestuffs.

Dead and unripe fibres in cotton from neps in the yarn : They do not bleach well and cause trouble in dyeing. They are dyed a deeper shade than ripe fibres with direct dyes. With basic dyes, they are sometimes dyed only in the interior, whereas ripe fibres are dyed uniformly with indigo, para red, etc., the neps appear as streaks or patches. Neps also cause trouble in mercerisation though it increases their affinity for dyestuff to some extent. Apart from the presence of neps in unspun cotton, they are also formed during spinning from the flabby, three-rolled hairs, which becomes roled up into tangled masses.

With Rayon, different lots of the same kind may have different affinities for the same dyestuff. The presence of traces of copper in cuprammonium silk is responsible for the tenderness in the bleaching of fabric.

(b) Defects due to faults in manufacture : Uneven dyeing may be due to any one of the following reasons :

(1) Faulty carbonisation resulting in the alteration of the affinity of wool for dyestuff.

(2) Structural damage of wool caused by excess of alkali or use of hard water while scouring.

(3) Damage caused by moulds or bacteria.

(4) Mineral oils or metals from machines.

(5) Differences in tightness of weaving of the fabric.

Sometime "Lists" appear in dyed goods. The lighter or darker coloured patches generally seen at the selvedge of the fabric are termed "Lists". They are attributed to the differences in tightness of weaving at the centre and edges of the fabric.

(c) Defects due to faulty scouring or bleaching : Cloudy or patch dyeing are often due to faulty scouring. Owing to the presence of residual impurities in incomplete scoured goods, they become impermeable in patches. These patchy portions cannot absorb completely the dye liquor resulting in different coloured patches in the fabric. The use of unsuitable water produces similar results by depositing calcium or magnesium soaps in the fabrics. An unevenly dyed fabric when examined is expected to be found to have been incompletely scoured which could not be completely permeable. The two pre-requisites of good dyeing are :

(1) The fabric should be completely permeable, and

(2) The fabric should be free from the impurities which are likely to affect the dyeing.

Soap or water which contains iron when used in scouring is likely to introduce impurities which would sadden the final colour of the fabric.

The use of unsuitable oils for lubricating the wool during spinning, weaving or knitting causes a lot of trouble in dyeing of the fabric afterwards. In knitting, a leather of soap and oil is generally used. Neatsfoot is more extensively used whereas the olive oil is the most suitable oil for a lather. No system of scouring with soap and water can remove the unsaponifiable oil which is generally present in neatsfoot to an extent of 10 per cent. Though it may not appear in the beginning but in the end this mineral oil will come up in the surface of the fabric and cause discolourations in white and uneven patches in the dyed fabric. Drying oils are still more offensive. The soap used in making the lather is never completely removed unless it is freely soluble in water of the scouring bath. The quantity of soap which is absorbed by wool, largely depends upon the solubility of soap. Residual soap is decomposed in the acid dye-batch with the liberation of fatty acids. These acids are likely to stick to the fibre causing patchy dyeing in the fabric. The action of calcium and magnesium soaps is also similar. Goods containing these soaps should be specially treated before dyeing. The fabrics should be first soaked in a warm solution of hyrochloric acid (1° TW) and washed with soft water for the removal of calcium or magnesium chloride. Then they are treated with dilute ammonia or sodium carbonate so that the liberated fatty acids are dissolved. The use of too little liquor or overcrowding in the scouring machine is also sometimes responsible for marks, lines, or streaks running irregularly in the longitudinal direction of the fabrics. In both the cases, the fabrics cannot change their positions and cause streakiness in dyeing.

Defects Due to Faulty Dyeing

Uneven shades may result from many reasons as follows :

(a) Rapid dyeing generally gives uneven shades and colours which are easily rubbed off. The fabrics should not enter the bath at too high a temperature. This holds good for all the fibres. The temperature of the bath should be such as to enable the fibre to develop the maximum affinity for the dyestuff. Rapid surface dyeing takes place when the fabrics enter the bath at too high a temperature whereas the liquor does not repenetrate the fibre.
(b) The temperature of the bath should not be raised too rapidly after entering the fabrics. In dyeing wool with acid dyestuffs, at least half an hour should elapse before the temperature of the bath is raised to the boiling point.
(c) The boiling should be continued for a sufficient length of time so as to allow of "equalisation".
(d) Acids, or dyestuff should not be suddenly added to the bath without lifting the fabrics. Solid chemicals instead of their solution also should not be added to the bath.
(e) With after chromed dye stuffs, potassium bichromate should not be added without allowing the bath to cool off.
(f) Care should be taken to see that the circulation of the dye liquor and the fabric is sufficient. This is more important in the case of heavy fabrics, or cops or cheeses where the penetration is difficult.
(g) The dyed fabrics should not be allowed to lie about without squeezing out the dye liquour and rinsing.
(h) The tension used in the fabrics in the open width in dyeing should be uniform.

Localised Stains, Patches or Spots: The above defects may be due to the following reasons :

(i) When the particles of undissolved dyestuff are present which stick rapidly to the surface of fabrics.
(ii) The use of hard water might precipitate the dye base on the fabric when dyed with basic dyestuffs. With the addition of soap to a dye-bath, hard water gives rise to calcium and magnesium soaps, which will result in incomplete scouring.
(iii) Uneven mordanting or when mordanted goods are allowed to lie about exposed to air and light before dyeing.
(iv) Mildew stains.
(v) Accidental splashes of dye liquor before or after the fabric is dyed.

Tenderness in Dyed Goods :

This is mainly confined to silk, rayon and cotton. It is generally caused by unremoved mineral acid.

Tendering of Weighted Silk :

Dyed silk is generally weighted and the dyeing and weighting processes are combined together. For weighting the silk, tannate of iron and a compound of tin are most commonly used. Weighting with tin cannot be carried out beyond a certain limit without damaging the fibre which appears as stains or tenderness on stirring. Silk is very sensititive to the action of hydrochloric acid and if there is any trace of stannic chloride left in the fabric, hydrochloric acid will be gradually produced by hydrolysis.

Tenderness in Dyed Cotton

Cotton is sometimes tendered by the action of unremoved traces of oxygen carried used when dyed with aniline black. Tendering may also be caused by sulphur dyestuffs, especially in the presence of an oxygen carrier. But all dyed goods are tendered, if there is a growth of moulds or bacteria.

Defects in Knitted Fabrics

Defects in knitted fabrics may be divided as follows:

(a) due to faults in yarn,
(b) due to faults in machine,
(c) due to careless handling of materials.

(a) The most common defects in yarns are cloudiness and cockling :

Cloudiness : These are thick and thin places caused by uneven yarn diameter.

Cockling : This is due to the use of harsh and wiry yarns, which may be improved upon in some cases by the proper lubrication or a suitable lubricant having even penetration.

(b) Defects due to faults in machine :

1. Tuck stitches : This is generally caused by :

(i) Too short a stitch.

(ii) Imperfect needles, bent or rough latches, damaged beards, etc.

(iii) When the timing of pressing and lending movement of bearded needle machine are incorrect.

(iv) Insufficient drawing-off power applied to fabrics, etc.

2. *Dropped Stitches :* These are generally due to :

(a) Incorrect setting of yarn guides.

(b) Imperfect needles.

(c) Improper adjustment of latch guards or brushes in latch needle machines.

3. *Uneven loops in a single course :* These are generally caused by :

(a) Wrong angle of stitch cam.

(b) Incorrect alignment of needles.

(c) Variation in tension on yarn.

The most common defect in fabrics knitted on multi feeder machines is the irregular loop length in different courses. Here also the defect is generally caused either by having more tension on some yarns than on others or by wrong adjustment of stitch cams.

(c) Defects due to careless handling the of the fabrics:

These defects are generally holes, stains or fuzziness, etc.

Holes are usually the result of the tying of unwidely knots. The ends should be neatly trimmed off after tying a good tiny knot. Rayon yarns should be kept spotlessly clean and free from oil stains and rough handling which cause fuzziness through the breaking of individual filaments.

Defects in Rayon Fabric : Common Weaving Faults

Barry places : These are mainly due to the erratic turning of the warp-beam, collars and pulley should be free from rust and frictional surfaces rubbed with French chalk.

Bright picks : Brightness is usually due to stretching beyond elastic limits. Resistance to unwinding from the shuttle may produce undue stretching.

Buttons : These are lumps on threads, the causes may be chafing with rough reed, worn healds, threads too low on shuttle race etc.

Cockled places : Made by stretching e.g., by strips of cloth or starting canvas wrapping round take up or other rollers.

Curls : Made when the shuttle fails to control the weft.

Regenerated rayon : The regenerated rayons are more affected by the same causes which bring tenderness in the cotton in bleaching. The use of too strong a bleaching bath rapidly brings about the degradation of viscose and cuprammonium Rayons. In a mixed fabric of cotton and viscose, the cotton might have been only slightly affected whereas the viscose would be degraded to such an extent so as to have no strength at all. In the case of cellulose acetate rayon which is a compound of cellulose, it is less affected than cotton. Strongly acid liquors also have a rapid tendering effect on regenerated rayons. Incomplete removal of acid after souring brings the degradation quickly owing to the formation of oxycellulose.

Detection of tenderness : The most common method of detecting the tenderness is to find the strength of the material under test and then to compare the results with the strength of the same fabric which is not in a tendered condition.

In the case of degradation in regenerated rayon, tensile strength in the wet state shows a greater percentage of decrease than the tensile strength in dry state. So it is most important to find the tensile strength in both dry and wet states.

If the copper number of the original material is known, tenderness can also be tested by the estimation of the copper number. Permanent photographic records are also very useful in the study of damaged goods.

A Nessler solution of a reduced causting soda content can detect the presence of oxycellulose in Rayon. The material under test and a piece of untendered fabric are boiled in this solution for a minute and then rinsed in dilute potassium iodide solution. The presence of oxycellulose will be indicated in case the sample under test becomes more deeply coloured than the tendered fabric.

Acetate Rayon : Damage to acetate rayon sometimes takes place by not ironing or other dry heat treatment. The tenderness is usually detected by the fact that the tendered fabric becomes stiff and the individual filaments are found to be plasticised together. But, in the case of mixed fabric of cotton and acetate rayon, the untendered portion is dyed normally whereas the tendered portion is scarcely dyed at all. So if a blue coloured fabric is to be over dyed with yellow, it will produced a green shade on the untendered portion, leaving behind the damaged portion to be blue as before.

Loss of Lustre : Cellulose loses its lustre partially or entirely when boiled for some time in water but this loss oflustre can be restored by treating the fabric with acetic acid, phenol, and solutions of ammonium sulpho-cyanide.

Acetate rayon, however, does not lose its lustre if it is boiled in solutions of sodium chloride, sodium sulphate and ammonium sulphate. A mixed fabric of cotton and acetate rayon can also be mercerised, as alkali of mercerising strength does not damage the acetate rayon.

Dyeing defects : Uneveness in dyed rayon fabric may be due to various causes. If a weft cop of different denier is used in the weaving of the fabric, a weft bar will be produced when the fabric is dyed even though the dyeing is perfectly even. This is an apparent difference to the eye, owing to difference in thickness and number of filaments in the composition of the yearn. This fault can be detected by a denier and filament test of the yarn from both light and dark places. Indiscriminate use of rayon yarn from various manufacturers also cause uneven dyeing in the fabric as the affinity of rayon for dyestuff may vary according to the source of the manufacture. This variability in dyeing affinity of the rayon is usually the cause of most uneven dyeing defects.

Before dyeing viscose rayon, it is better to determine which dyestuffs will evenly dye the fabric since viscose rayon is liable to variation in dyeing affinity.

The chemical composition of acetate rayon is usually

kept very constant so that dyeing faults are less frequent than in the case of viscose.

Capillarity Tests

When threads of viscose having the same dyeing affinity are suspended in such a way that the lower ends of the threads are dipped into solutions of various dyestuffs, it would be noticed that some of the threads are coloured to a greater height than the rest. This will indicate that those dyestuffs which rise the least will dye evenly whereas those that rise to the greatest height will dye unevenly.

19

TESTING OF DYED MATERIALS UNDER VARIOUS TESTING PROCEDURES

The outstandingly important prosperity of a dyed material is the fastness of the shade. It has not been easy to find a means of expressing fastness in unequivocal terms, because any simple and universally applicable assessment of alteration of colour must be based on subjective judgement. A number of tests are necessary to cover all the important properties of any one dye because good fastness to one influence is not necessarily accompanied by equal fastness to exposure to other conditions. Tests may be divided into those of consumer significance, such as light, wash, and salt-water staining, and those concerning only the producer, such as fastness to cross-dyeing or unshrinkable treatments.

There was, at one time, an unfortunate lack of uniformity in methods of testing and expression of fastness. It was for the purpose of bringing about greater uniformity that the Fastness Tests Committee of the Society of Dyers and Colourists was set up in 1927, and it did much to co-ordinate methods of testing within the field of the Society's influence. There was, however, little international uniformity and in 1947 the International Organisation for Standardisation (I.S.O.) formed a colour-fastness sub-committee. On account of its international nature the I.S.O. cannot impose its decisions, and the term I.S.O. Recommendation is therefore, used for a test upon which final agreement has been reached.

The fastness of a dye is related to the depth of the shade. If a dyed sample be exposed to light, the number of molecules which will be modified during a specified period will be the same, irrespective of the percentage of dye of the fibre. The proportion of those present which are decomposed is therefore, much greater in a pale shade. It is not possible to standardize dyeings for testing on a percentage basis because of the variations in strengths of dyestuffs. A range of standard depths in 18 hues has been prepared and is available (from the Society of Dyers and Colourists), known as the 1/1 standard depth series. This should be matched, as closely as possible, when preparing the sample for testing. The use of 1/1 range is recommended whenever possible, but the 2/1 standard depth is twice as strong and weaker ones available are the 1/3, 1/6, 1/12, and 1/25, respectively.

Assessment of fastness involves a visual determination of either change in shade or staining of an adjacent material. An entirely subjective description such as good, moderate, or poor has no value. The first effort to introduce greater precision was the pass/fail tests. The specimen was subjected to a number of tests of increasing severity, and passed as long as no alteration was apparent, the fastness being expressed as the number of the most severe test which the dyeing passed. The method was tedious in use and its sensitivity was poor. It is not surprising, therefore, that the use of dyed standards found favour.

A series of dyes was specified as corresponding with grades of fastness ranging from 1 to 8 for light and 1 to 5 for other tests. Materials dyed to specified strengths with these colours were tested simultaneously with the sample, and the standard which showed the same change of shade or staining indicated the grading. The method was not entirely satisfactory on account of the large number of standard samples which became necessary.

Most of the difficulties are overcome by the use of grey scales against which it has been found possible to compare loss of colour or staining of any hue irrespective of depth. It is necessary to use two scales, one for assessing change in colour and the other for staining.

Determination of Light Fastness

Exposure of daylight is a test of the behaviour of the sample under actual conditions of use, but takes a long time. Sources of light with spectral distribution as near as possible to daylight give quicker results but do not necessarily provide a precise prediction at behaviour during use.

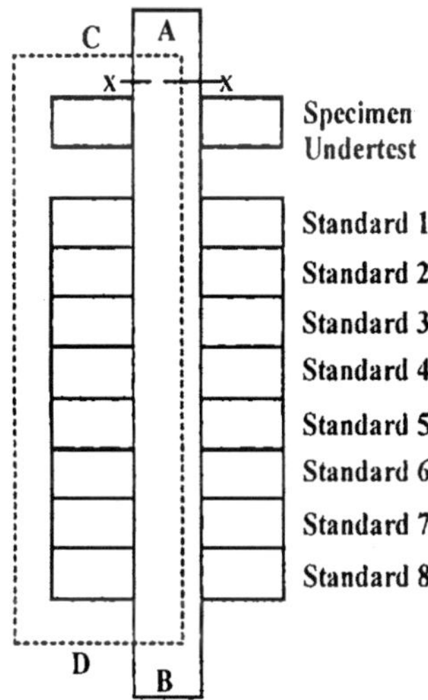

Fig. 1(a). Mounting of specimen and standards for testing fastness to light. (I.S.O. Recommendation R102)

The specimens are mounted in a frame facing south in the Northern Hemisphere and north in the Southern Hemisphere at an angle approximately equal to the latitude of the place where the exposure is made. The rack should be placed so that no shadows fall on it, and the specimens must be protected against rain by a glass sheet not less than 5 cm away and supported in such a manner that the samples are well ventilated. One-half of each sample and standard is covered with an opaque sheet of cardboard or aluminium, leaving the other half exposed.

The specimen and control remain in the frame 24 hours per day until sufficient fading has occurred for a good comparison to be made with the standards. The ideal

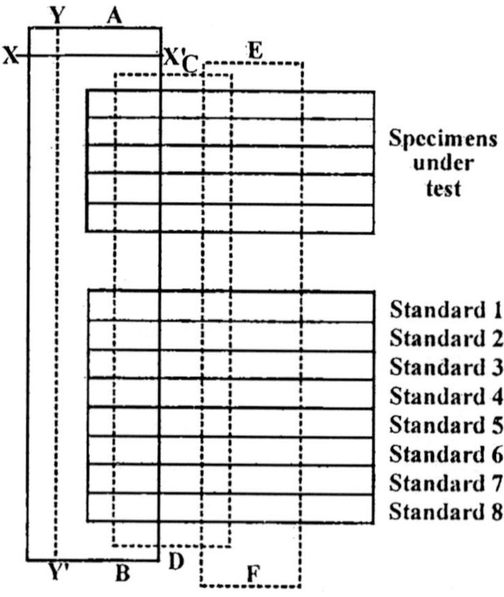

Fig. 1(b). Mounting of a number of specimens and standards for testing fastness to light.

way of carrying out the test is to mount the standards and specimen as shown with a hinged opaque cover, and expose until the contrast between the covered and exposed portions of the specimen is equivalent to grade 4 on the change of colour grey scale. At this stage cover up a second previously—exposed portion of the patterns with an opaque screen CD and continue to expose until the contrast between the continuously illuminated and completely shielded portions are equal to grade 3 on the grey scale, unless standard 7 has previously faded to grade 4 when the test may be terminated. The specimen and standards will show three zones with the unexposed are in the middle. The zones are compared with the standards and, if the two degrees of fading on the specimen do not correspond with the same standard, the fastness would be the mean between the two.

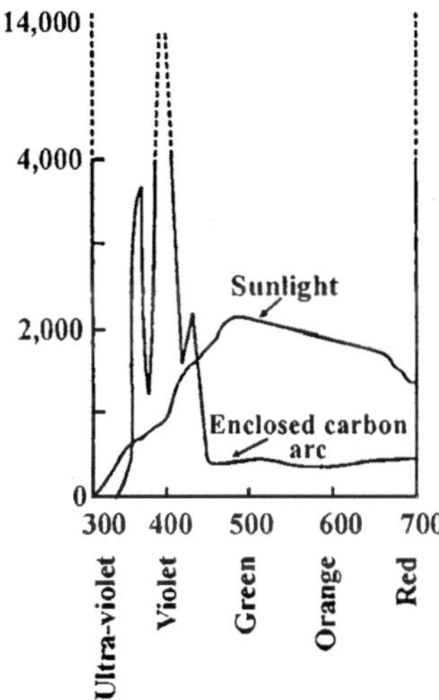

Fig. 1(c). Spectral energy distribution of daylight and carbon arc.

If the relative humidity of the surrounding atmosphere remains constant. There is a tendency for dyes to fade faster at higher temperatures.

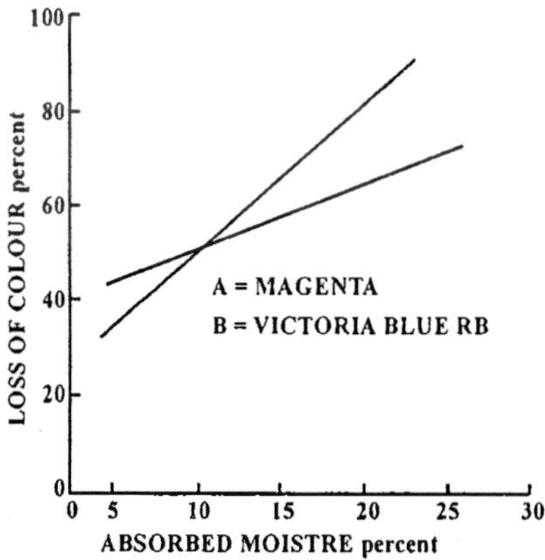

Fig. 1(d). Effect of moisture on light fastness.

The effect is much more marked with some dyes than with others. Humidity however, has a greater influence on the rate of fading than temperature.

In the graphs below, the action of light on Magenta and Victoria Blue RB are related to different moisture contents of the substrate corresponding with varying relative humidities of the surrounding atmosphere. It is interesting

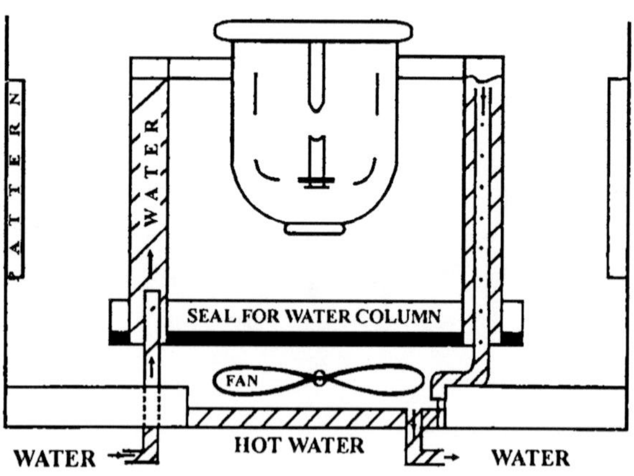

Fig. 1(e). Constant humidity enclosed arc fading lamp.

to note that with 5 per cent absorbed moisture (in equilibrium with atmosphere at 50 per cent relative humidity), Victoria Blue RB fades faster than Maganta, but the reverse is the case with 20 per cent absorbed moisture (100 per cent relative humidity).

Fastness to Washing

This is of great importance to the consumer, and there several washing tests which are applied according to the purpose for which the material is intended. They are proposed by I.S.O. and are carried out in the following manner.

A specimen measuring 10 cm × 4 cm of the material to be tested is cut out. Yarn is knitted into a fabric from which a piece of the same dimensions can be obtained, and loose fibre is compressed into a sheet measuring 10 cm × 4 cm.

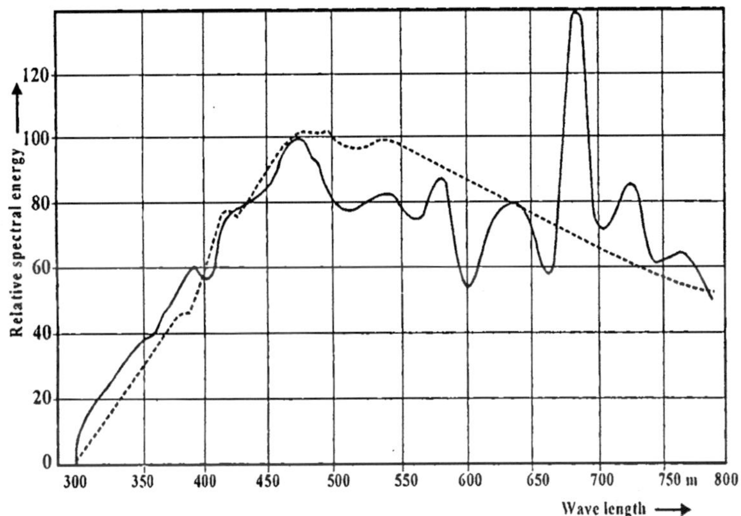

Fig. 1(f).

_____ Xenon are Xe 1500, filtered through the glass-water filter in the exnotest.

------- Natural radiation, i.e. radiation of sun and sky on horizontal plain in medium latitudes.

The specimen to be tested is placed between two pieces of undyed fabric measuring 5 cm by 4 cm and the three pieces are held together by stitching round the edges, leaving 5 cm × 4 cm exposed. In the case of the loose fibre the compressed mass is held in place by sewing it between pieces of cloth measuring 10 cm × 4 cm.

The composition of one of the colourless materials enclosing the specimen will be the same as the dyed sample and the other will be as indicated below :

If the first piece is	The second piece will be
Cotton	Wool
Wool	Cotton
If the first piece is	The second piece will be
Cotton	Wool
Wool	Cotton

Silk	Wool
Linen	Wool
Viscose	Wool
Cellulose acetate	Viscose
Polyamide	Wool or viscose
Polyester	Wool or cotton
Acrylic	Wool or cotton

A solution is made containing 5g per litre of soap conforming with the following specification :

Free alkali calculated as Na_2CO_3	0.3 per cent maximum
Free alkali calculated as NaOH	0.1 per cent maximum
Total fatty matter	85 per cent minimum
Titre of mixed fatty acids contained in the soap	30°C (86° maximum)

Procedure-1

The composite specimen is treated in a wash wheel or an equivalent apparatus at 40° ± 2°C for 30 minutes, using sufficient of the soap solution to give a liquor ratio of 50 : 1.

Procedure-2

The composite sample is treated in a wash wheel or equivalent apparatus for 45 minutes as 50° ± 2°C, using sufficient of the soap solution to give a liquor ratio of 50 : 1.

Procedure-3

To the soap solution, 2g per litre of anhydrous sodium carbonate are added. The composite sample is then treated in a wash wheel or equivalent apparatus at 60° ± 2°C for 30 minutes in sufficient of the above solution to give a liquor ratio of 50:1.

After treatment the composite samples are in every case rinsed twice in cold distilled water and then for 10 minutes in cold running tap water. After squeezing, the stitching is removed on the two long sides and one short side, leaving the dyed specimen and the undyed materials sewn together only along one short side. The pieces are opened out and dried in air at a temperature not above 60°C (140°F).

The change in colour of the uncovered portion of the specimen is assessed with grey scale no. 1 and staining of the undyed materials with grey scale No. 2.

Procedure-4

The dyed specimen, measuring 10cm × 4 cm as before is sewn together with two pieces measuring 5cm × 4 cm. One piece is of the same material as the specimen.

If the first piece be	The second piece must be
Cotton	Viscose
Linen	Viscose
Viscose	Cotton
Cellulose acetate	Viscose
Polyamide	Viscose or cotton
Polyester	Viscose or cotton
Acrylic	Viscose or cotton.

The composite sample is placed in a wash-wheel container together with ten balls and sufficient solution containing 5 g of soap and 2g of anhydrous sodium carbonate per litre to give a liquor ratio of 50:1. The test is carried out at 95° ± 2°C for a period of 30 minutes.

The rinsing, drying, and assessment of the sample are carried out in exactly the same manner as was the case with tests 1 to 3.

Procedure-5

The composite sample is prepared as in the previous cases with the exception that :

If the first piece be	The second piece must be-
Cotton	Viscose
Linen	Viscose
Viscose	Cotton
Cellulose acetate (triacetate only)	Viscose
Polyamide	Viscose or cotton
Polyester	Viscose or cotton
Acrylic	Viscose or cotton

A 50:1 liquor ratio is used with a solution containing 5g of soap and 2 g of anhydrous sodium carbonate per litre. The test is carried out at 95° ± 2°C for a period of 4 hours. The rinsing, drying, and assessment are carried out as in the previous tests.

Fastness to Perspiration

Unless care is taken in the selection of suitable dyes, those parts of a garment which come into contact with the body where perspiration is heavy may suffer serious local discolouration. Two artificial perspiration solutions are made up as follows :

	Solution	
	A	B
1-Histidine mono-hydrochloride mono-hydrate	0.5 g	0.5 g
Sodium chloride	5.0 g	5.0 g
Disodium hydrogen orthophosphate ($Na_2HPO_4.2H_2O$)	2.5 g	2.2 g
Volume in distilled water	1000ml	1000 ml
pH (Adjusted with N/10 NaOH)	8	5.5

The material to be tested is placed between undyed pieces, one of which is of the same fibre and the other of an alternative composition, according to the list quoted in the I.S.O. recommended hand-washing test. The pieces are sewn along one side to form a composition specimen, and this is thoroughly wetted out in 50 times its weight of the artificial perspiration solution (A). It is allowed to remain in the solution for 30 minutes at room temperature, after which the liquid is poured off. The specimen is then placed between two glass plates pressed together with a force equivalent to 4.5 kg (10 lb) and allowed to stand in an over at 37° ± 2°C for 4 hours. After this the dyed sample and undyed cloths are separated and they are all dried at a temperature not exceeding 60°C; the procedure is repeated with solution (B). The changes of colour and staining are than assessed with grey scales for each test.

An alternative procedure is to use the method recommended in the second report of the Society of Dyers and Colourists, Fastness Committee. Two composite specimens are wetted out in solution (A) and (B) employing a 20:1 liquor ratio by immersion at room temperature for 30 minutes. Each sample is then laid out in a glass dish with a smooth flat botton, and a glass plate weighing about 50g is laid over the top and the whole is pressed to remove air bubbles and, after standing for 15 minutes, the solution is poured off without removing the plate. With no further disturbance the apparatus is heated to 37° ± 2° in an oven for 4 hours.

It must be borne in mind that some dyes are salt-sensitive, the change in shade being caused by the sodium chloride. Such dyestuffs can be discoloured when splashed with sea water, but salt staining is usually removed by rinsing in fresh water. Perspiration can also remove the metal from some copper after-treated direct dyes resulting in loss of wet and light fastness.

Fastness to Chlorinated Water

This test is intended to assess the resistance of dyed materials to concentrations of active chlorine, similar to those which may be present in swimming baths. A solution containing 20 mg of active chlorine per litre is prepared by first diluting 13.3 ml of sodium hypochlorite, containing 150 g per litre of active chlorine to 1000 ml, within distilled water, and then adding 10 ml of this 1000 ml of a solution buffered to 8.5 with –

5.1 ml of 0.1 M sodium hydroxide solution

0.0144 g of potassium chloride per litre

0.1123 g of boric acid

The pH should be checked, also the available chlorine determined by titration, and adjustments to both made, if necessary. The sample is moistened in distilled water, squeezed, and immersed in 100 times its weight of the chlorine solution at room temperature for 4 hours. It is then rinsed thoroughly, dried at room temperature, and any change of colour is assessed with the grey scale.

The tests which have been described are connected with those properties of importance to the consumer. There are also a number of fastnesses which are of interest to the producer, and have particular application to dyed yarns incorporated during knitting or weaving to create a pattern which must withstand subsequent bleaching or dyeing.

Fastness to Hypochlorite Bleaching

The sample is wetted out and squeezed. It is then immersed in a solution of sodium hypochlorite containing 2 g per litre of available chlorine, and buffered to pH 11 ± 0.2 with 10 g per litre of sodium carbonate. The liquor ratio should be 50 : 1 and the material to be tested is immersed at 20° ± 2°C for 1 hour under conditions such that there is no exposure to direct sunlight. The sample is then rinsed in cold running water and immersed at room temperature for 10 minutes in a solution containing 2.5 ml per litre of 30 per cent (wt./vol.) hydrogen peroxide, or alternatively, 5 g of sodium bisulphite ($NaHSO_3$) per litre. After drying at a temperature not exceeding 60°C, any change of colour is assessed against the grey scale.

Fastness to Peroxide Bleaching

The specimen is sewn between one piece of cloth of the same fibre and another of :

(a) cotton, if the sample be wool, silk, viscose rayon or linen.

(b) Viscose rayon if the sample be cotton or acetate rayon. It is then rolled up and placed in a test tube and covered with the appropriate peroxide liquor. The composition of the liquors and the conditions of the tests are set-out in table 1 below.

Table 1 : Composition of Bleach bath and Conditions of use.

Starting bath (per litre of distilled water)	Bath 1 (for natural and regenerated cellulose)	Bath 2(h)	Bath 3 (for wool and acetate rayon)	Bath 4 (for silk)
Hydrogen peroxide (a), ml	5	—	20	20
Sodium peroxide (b) g	—	3	—	—
Sodium silicate (c), ml	5	5	—	5
Sodium pyrophosphate (d) g	—	—	5	—
Magnesium chloride (e) g	0.1	0.1	—	0.1
pH, initial value ± 0.2(f)	10.5	11.5	9.3 (g)	10
Temperature, °C ± 2°	90	80	50	70
Temperature, °F ± 4°	194	176	122	158
Duration of treatment, hr	1	1	2	2
Liquor ratio	30 : 1	30 : 1	30 : 1	30 : 1

(a) 30.4 per cent H_2O_2 (wt./vol.)

(b) 100 per cent Na_2O_2

(c) About 26 per cent SiO_2 and 10 per cent Na_2 (sp gr = 1.32, 35° Be)

(d) Na_4P_2, $10H_2O$

(e) $MgCl_2$ $6H_2O$

(f) Adjust by addition of NaOH solution if necessary.

(g) The pH of the bath at the end of the test must not be less than 9.0.

(h) Bath 2 is included primarily to meet the requirements of some continental countries.

The degree of staining and alteration in colour are assessed with grey scales.

Fastness to chlorite bleaching

The Dyed material is sewn between two piecees of cloth of the same fibre. The test solution contains 1 gm of sodium chlorite (80 per cent) per litre and before the test, sufficient acetic acid is added to bring the pH to 3.5. The specimen is wetted out in the solution and subsequently immersed in it at 80° ± 2°C (176° ± 4°F) for 1 hour, using a liquer ratio of 50 : 1. The sample is then washed in cold running water for 10 minutes and dried at-temperature below 60°C (140F).

Fastness to Chlorination

The test is designed for yarns or goods which will, at a later stage of manufacture, receive unshrinkable treatment . A specemen of the dyed material measuring 7.5 × 5 cm is recommended. If it be desirable to ascertain the degree of staining of other fibres, a few stitches of the appropriate undyed yarns are sewn in to the fabric at approximately 1-cm intervals. The sample is immersed for 10 minutes at room temperature in 25 times it's weight of a solution of hydrochlortic acid containing 6 ml (1.16 sp. gn) per litre of the acid; then an equal volume of sodium hypochlorite solution containing 1g of available chlorine per litre is added, and treatment is continued in a cold liqnor for another 10 minutes. The sample is next rinsed thoroughly in cold running water and subsequently dechlorenated in a 3g per litre sodium sulphite (Na_2 SO_3 $7H_2O$) solution at 35° to AO° (95° to 104°F), using a 50 : 1 liqnor ratio for a period of 10 minutes. After rensing and drying at temperatures not exceeding 60°C (140°F), assessments of alteration of colour and staining are made with the aid the appropriate grey scales.

Fastness to Cross-dyeing

The purpose is to test yarn intended for use with wool, and which should withstand dyeing by all the methods that may be used for the protein fibre. The yarns should be knitted into fabrics for the preparation of the samples. A piece of the dyed material measuring 10 cm × 4 cm is placed between two pieces of undyed cloth and sewn round the edges. One undyed pieces should be of the same fibre as that of the sample undergoing test, and the other selected as follows :

Fibre under test	Composition of alternates undyed piece
Cotton	Wool
Wool	Cotton
Silk	Wool
Linen	Wool
Viscose rayon	Wool
Cellulose acetate	Wool
Polyamide fibre	Wool
Polyester fibre	Wool
Acrylic fibre	Wool

The specimen is then tested in the following different way corresponding with the methods by which wool may be dyed, the liquor ratio always being 50 : 1 and the percentage being based on the weights of the protein and polyamide fibre in the composite sample.

(1) Neutral Cross-dyeing

Enter the composite specimen into a liquor containing 20 per cent of the sodium sulphate crystals. The temperature is raised to 92° ± 2°C over a period of 30 minutes, and maintained for a further 90 minutes.

(2) Acetic acid cross-dyeing

The procedure is as in (1) except that the liquor contains 5 per cent of acetic acid (30 per cent) and 20 per cent of sodium sulphate crystals.

(3) Sulphuric acid cross-dyeing

The procedure is similar, but a solution containing 4 per cent of sulphuric acid (sp gr. 1.84) and 20 per cent of sodium suslphate crystals is used.

(4) Acetic acid, chrome cross-dyeing

The composite specimen is entered into a bath containing 20 per cent of sodium sulphate crystals and 5 per cent of acetic acid (30 per cent). The temperature is raised to 98° ± 2°C over a period of 30 minutes. At which it is then maintained for a further 30 minutes, and then 2 per cent of potassium dichromate is added and the temperature is maintained at 98° ± 2°C for another 60 minutes.

(5) Sulphuric acid, chrome cross dyeing

The specimen is entered into a liquor made up with 20 per cent of sodium sulphate crystals and 3 per cent of acetic acid. The temperature is brought up 98°C ± 2°C over a period of 30 minutes, and after a further 30 minutes, 2 per cent of sulphuric acid (sp gr 1.84) is added. Then after another 15 minutes, 2 per cent of potassium dischromate is added. The temperature is now kept at 98° ± 2°C for 1 hour, after which the test is complete.

Whichever method is used the samples are rinsed in cold running water, dried at a heat not greater than 60°C, and the change of colour and staining are assessed with the grey scales.

Fastness to Mercerizing

If the sample to be tested be fabric, a portion measuring 10 cm × 10 cm is sewn round its edges to a piece of undyed bleached cotton of the same size. The composite specimen is then fastened firmly, but without excessive tension, to a frame with the coloured portion uppermost. Dyed yarn is wound on to a rigid frame, without excessive tension, with the parallel strands arranged closely together, until an area of 10 cm × 10 cm is covered. A piece of undyed bleached cotton cloth is then attached by sewing, leaving the coloured yarn uppermost.

The specimen, on the frame, is immersed in a solution containing 300 g of sodium hydroxide (NaOH) per litre at 20° ± 2°C for 5 minutes. It is then rinsed, whilst still on the frame, with a litre of water at 70° ± 2° C for 1 minute, and finally in runing cold water for 5 minutes. The sample is next removed from the frame, and the residual alkali is neutralized by immersion for 5 minutes in 50 times it's weight of sulphuric acid of a concentration of 5 ml per litre. After rinsing the acid out in cold running water, the material is dried, the temperature not exceeding 60°C, the dyed and undyed portions having been separated except at the stitching along one side.

Many colours show an apparent increase in depth on mercerizing. In such cases the rating of 5 is distinguished with an asterisk (5*) or, if the hue alters without loss of depth, the grey scale grading is used to indicate, not the loss of colour, but the extent of the alteration (3-4 redder*).

There are many other fastness tests and for further information the second edition of the Society of Dyers and Colourists, publication Standard methods for the determination of the colour fastness of textiles should be consulted. The only process for which no test has been published is dealgination, a treatment used almost invariably with men's socks of wool or wool and cotton. The method found satisfactory by various tests has been to attach white cotton to dyed wool or the reverse, and treat at 60°C for 30 minutes in a solution containing 2.5 g per litre of tetrasodium pyrophosphate.

Identification of Dyes

It is often necessary to ascertain what dye or class of dyestuff has been used on a pattern before goods can be dyed in bulk to match. An exact identification is often difficult because a mixture of dyes will nearly always be used in practice.

In the first place, it is helpful if the class rather than the individual dye can be determined, and in very many instances this information is sufficient. In the case of wool and silk the probable classes are basic dyes, acid dyes, and mordant or premertallized dyestuffs.

The basic dyes differ from all others because they are extracted quite easily by boiling with acetic acid or with alcohol. However, if these reagents do not remove a substantial portion of the dye from the fibre, the next test is to boil the coloured material in dilute ammonia (1 ml of

0.88 ammonia per 100 ml of water) for 1 to 2 minutes in the presence of white cotton.

If the solution is distinctly coloured and the cotton remains unstained, the presence of an acid dye is indicated. This can be confirmed by boiling with 40 per cent of Glauber's salt when a considerable proportion of molecularly dispersed acid dyes are stripped, but little effect is apparent with the fast acid dyes. If the cotton is stained the presence of a substantive dye or a structurally analogous neutral dyeing acid colour is indicated. It must be emphasized, however, that there are many acid dyes which are strongly absorbed by wool in a neutral liquor and leave the cellulose unstained.

If, on boiling with ammonia, virtually no colour migrates into the liquor, mordanted or premetallized dye is probably present. This is confirmed by testing for the presence of appropriate metallic ions. To do this, the material must be ashed, and the method recommended by Clayton is to heat about 1 sq. in, if possible, in a platinum or silica dish until charring has taken place. The dish is then cooled and the residue is covered with a hot saturated solution of sodium nitrate, the water is evaporated, and heating is continued until and the carbon has burned away. The ash is extracted with water or dissolved with the aid of a little hydrochloric acid, and the usual qualitative tests for metallic ions are applied.

When the general group to which the dye or dyes belong has been determined, much useful information about the nature of the chemical constitution can be gained by studying the action or Formosul G. This regent is prepared by dissolving 75 g of Formosul in 75 ml of hot water, and then diluting with 75 ml of cold water and 50 g of mono or di-ethylene glycol. The specimen is boiled with the reagent for 1 to 2 minutes, and its behaviour can be compared with the information given in Clayton's tables.

With regard to cellulosic fibres, the simplest test for the dyes which will probably be present, ignoring unlikely ones, such as acid or basic dyes, is to boil for 1 to 2 minutes with a 5 per cent caustic soda solution. If a considerable quantity of colour is stripped, a substantive dye is indicated and it will probably be an azo dye, if the colour of the fresh sample is destroyed by boiling for 2 minutes with Formosul G. If, on the other hand, very little or no dye is stripped, the inference is that the material has been dyed with a sulphur, azoic, vat, or reactive dye. When boiled for 1 to 2 minutes with Formosul G and a few drops of sodium hydroxide solution, azoic or reactive dyes are virtually stripped. They can, however, be distinguished because only reactive dyes and certain phthalocyanine derivatives, usually recongnizable by their distinctive colour, well withstand the following treatment.

Boil for four minutes, in succession, in the following reagent, rinsing in distilled water after each treatment :

(i) Glacial acetic/acid and ethanal (1 :1 vol/vol.).

(ii) 1 percent ammonia solution. (sp. gr 0.88, 10 ml/litre.

(iii) Dimethyl Formamide and water (1:1 vol/vol.).

(iv) Dimethylformamide.

If the pattern, on boiling with Formosul G and alkali is decolourized but the colour is restored by rinsing in hot water, the presence of a sulphur dye is suggested. Confirmation is provided by boiling for half a minute with 16 per cent hydrochloric acid, cooling, and adding a little zinc when hydrogen sulphide will be evolved, detectable by the black colour produced on filter paper moistened will lead acetate solution. As an alternative, the sulphur can be reduced to hydrogen sulphide with an acidified solution of stannous chloride.

If the dyeing changes shade when treated with alkaline Formosul G, but the colour is restored on exposure to air, or when treated with hydrogen peroxide or ammonium or potassium persulphate, and it does not respond to the test for a sulphur colour, it will be a vat dye. Some sodium salts of leuco compounds differ little in colour from the oxidized pigment, but in such cases the free leuco compound, obtained by reduction with a boiling solution containing equal proportions of Formosul and acetic acid, will often show a more marked change of colour which is restored on oxidation or exposure to air. According to Bradley and Derrett-Smith, if small specimen of the dyed sample be wrapped in a piece of bleached linen or cotton measuring about 1 sq inch and tightly bound into a role with undyed sewing thread and boiled for a few seconds in alkaline hydrosulphite, the leuco compound marks off on to the white material and the colour is developed on oxidation. Although both vat and sulphur dyes mark off, the latter can easily be recognized by the hydrogen sulphide test.

When the class of the dye has been determined, it is desirable, if possible, to identify the actual dyestuff. A simple but not very reliable method is to place cuttings in a spotting plate of white porcetain with series of small depression about ½ inch in diameter, arranged in parallel rows. Three or four or more cuttings of the same fibre dyed with known colours of similar shade from the corresponding group are placed in parallel rows. The sample and cuttings of known dyestaffs are then spotted with drops of reagents, such as concentrated sulphuric acid, nitric acid, hydrochloric acid, 20 per cent sodium hydroxide solution, acedified stannous/chloride and alkaline hydrosulphite. The colourations which result are compared and if any of the series shows the same reaction as the sample, this is taken as evidence of identity. The method is entirely empirical, not very reliable and of no value unless the sample has been dyed with only one dyestuff. It is often more useful as a source of negetive information when it is desirable to prove that two batches were not dyed in the same way.

Chromatography

A very promising field for the examination of dyed materials which has been receiving much attention is chromatography. The method has been used, in a crude manner, for indicating whether dyes are homogeneous or mixtures, for a long time. Some of the dye is stripped from the material with boiling alkali, pyridine, ethylene diamine, or other solvent which does not cause decomposition, and then converted to aqueous solution, into which strips of filter paper are suspended so that the lower end makes contact with the surface and the remainder (6 to 12 in.) is not immersed. If the dye be a mixture, the different components will rise to various levels in the strip. It must be borne in mind that the separation is only effective when the dyes have no marked affinity for the cellulose of which the filter paper is composed.

A more efficient separation is described by Clayton. A tube is filled with an absorbent material, one of the most satisfactory being a specially prepared alumina. The dye, in solution, is introduced through a dropping funnel at the top of the column, and is initially absorbed by the top layer of aluminia. A suitable solvent is then allowed to percolate downwards, accelerated by the application of suction if necessary. After the solvent, often referred to as the eluant, has flown through the column for a suitable time, often 2 to 3 hours, the components will be separated into clearly defined zones, A, B and C in Fig. 1(g). The column can then be broken up, the bands separated and the dyes isolated from them or, alternatively, it may be eluted with suitable solvents which will carry the zones, one after another, into the receptacle at the bottom of the column.

Paper chromatography has now, to great extent, replaced alumina or similar absorbent columns. An advantage is that much small quantities can be used. The apparatus is illustrated diagrammatically in Fig. 1(i). The sheet of filter paper is suspended so that its bottom dips into a vessel containing the eluent and the upper end is attached by clips to a frame.

The whole can be placed in a rectangular glass vessel with ground edges at its open end, so that an airtight seal can be formed when covered with a glass plate. A spot of dye solution is placed on the starting line and it will rise to a height, where it formes a stationary coloured patch at a distance D-d from the height D to which the solvent rises. The factor R_f, which should be constant for the same compound at the same temperature with the same solvent, is the ratio of the distance travelled by the spot to the distance travelled by the eluent or d/D/. But it is suggested that the R_f value is not very reliable for identification unless supported by actual chromatograms of known dyes. Chromatograms may be prepared with either an ascending or a descending eluent.

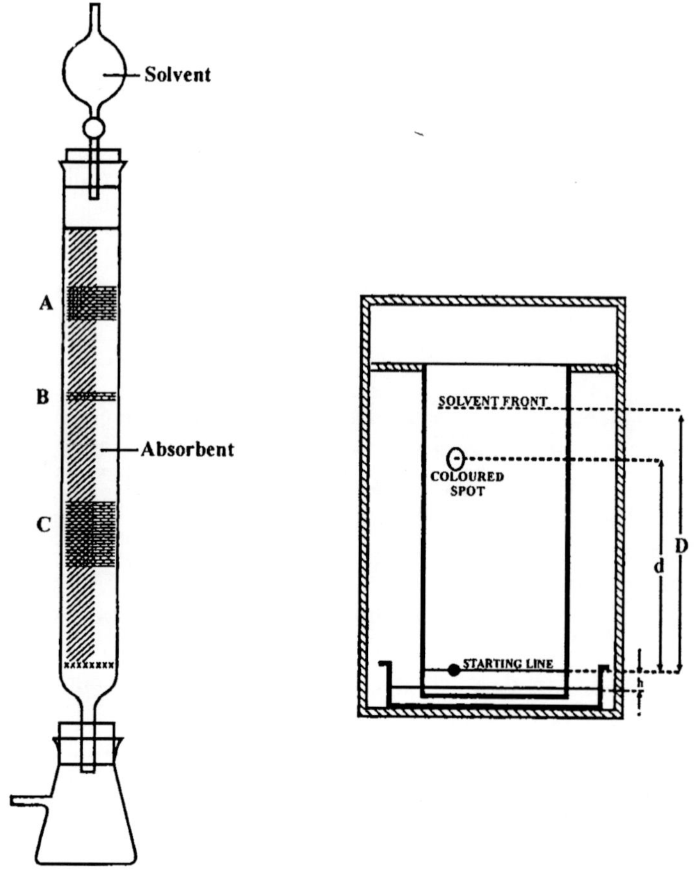

Fig. 1(h) Simple apparatus for chromatoraphic separation.

Fig. 1(i). Diagram of apparatus for paper chromatography.

Classical absorption chromatography relied upon the strength of the conflict between molecular attraction of the substance under examination for the absorbent and the tendency for the eluent to dissolve and remove it. Alumina or paper was the absorbent and usually only one solvent, very often water, was percolated. Partition chromatography is a refinement of this procedure. The absorbent in this case is a stationary liquid supported on a liquid subtrate such or filter paper. The rate of travel of the dye depends upon partition between the stationary liquid and the eluent. An eluent mixture is made up of :

Cyclohexane-200 ml, Glacial acetic acid 192 ml, water 8 ml.

This mixture is shaken in a separating funnel and allowed to stand overnight. It will separate into a concentrated acetic acid phase in equilibrium with cyclohexane, and a layer composed essentially of cyclohexane. To prepare a descending chromatogram the lower layer in the separating funnel is run off the next day poured into the bottom of the chromatography tank. The paper, which has previously been spotted with the dyes, is placed in position, the lower edge not actually making contact with the liquid. It is allowed to remain in this position overnight so that by absorption of the vapour a layer of acetic acid in equilibrium with cyclohexane can be built up on the paper. Then the next morning, the upper layer from the separating funnel is run into the trough at the top of the tank, and a period of 4 to 6 hours is allowed for completion of the chromatogram. It must be borne in mind that direct dyes do not respond well to paper chromatography because the affinity which they have for cellulose makes the colour relatively immobile. An eluent composed of :

Benzyl alcohol, 90 ml.
Dimethyl formamide, 60 ml.
Water, 60 ml.

Has proved satisfactory for direct dyes, because the dimethylformamide is a powerful hydrogen bonding agent.

If a dyed material is to be examined the dyestuffs may be extracted in the following manner.

Direct Dyes

Boil for about 2 minutes in a mixture composed of 2 parts of dimethyl formamide and 1 part of water. The extracted dye solution is concentrated by evaporation under reduced pressure. It must be borne in mind that dimethylformamide is toxic and an acute irritant.

Acid Dyes

It is usually possible to strip sufficient of the dye by boiling in a 10 per cent solution of 0.88 ammonia. The extract can be concentrated atmospheric pressure and the dye, if somewhat insoluble, retained in solution by the addition of a drop of 1 per cent sodium carbonate solution.

Metal Complex Dyes

These can be stripped by boiling in a mixture of 2 parts or dimethylformamide and 1 of water for 15 to 30 minutes, and then concentrated under reduced pressure.

Chromatography Applied to Vat Dyes :

This is a method specially designed for the chromatographic examination of vat dyes. Owing to the insolubility in virtually all solvents of the pigment form, the separation must be carried out in the leuco state. The reduced vat dyes, however, have such a marked affinity for cellulose that paper chromatography must be ruled out, but strips of material made of hydrophobic synthetic fibres, of which polyesters are the best, may be substituted. According to the report of Rao, Shah and Venkataraman they were able to separate vat dyes on paper or on powdered cellulose, if they used sodium hydrosulphate and tetraethylenepentamine instead of sodium hydroxide as the vatting mixture. Klingsberg (loc. cit.) describes an apparatus with which chromatograms of vat dyes can be obtained on strips of paper using as the eluent 10 g of sodium hydrosulphite dissolved in 100 ml of a 10 per cent solution of tetraethylene-pentamine.

A. 60-ml dropping funnel
B. Stopcock
C. Rubber stopper
D. Glass tube
E. 50-ml beaker
F. Dye specimen
G. Paper strip
H. Glass rod support

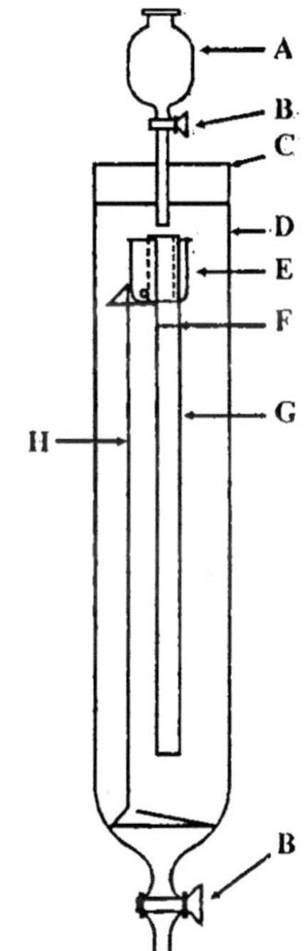

Fig. 1(j). Apparatus for paper chromatography under nitrogen.

The whole is enclosed in a wide glass tube (D) which can be evacuated and filled with nitrogen to exclude the possibility of oxidation of the leuco compound to the insoluble pigment during the preparation of the chromatogram. A descending chromatogram is prepared by suspending the upper end of the strip of paper in a small beaker into which the eluent is delivered from the stoppered funnel.

20

TESTING OF COLOUR

Theory of spectrum and substractive Primaries

Colour is a very important aspect of Textile and testing of colour is an integral part of it. Electromagnetic radiations range from X-rays with a wavelength of 10^{-11} cm, to radio waves, varying from 10 to 10^4 cm. As illustrated in Figure below 1, visible light constitutes only a very narrow band in this spectrum, lying between 4×10^{-5} cm and 7.2×10^{-5} cm. Wavelengths are also expressed in Angstrom units (A equals 10^{-8} cm or 10^{-10} metre), and in millimicrons (1 mm 10^{-7} cm or 10^{-10} metre).

γ-rays	X-rays	Ultraviolet	Visible	Infrared	Short waves	radio
10^{-11}	10^{-9}	10^{-7}	10^{-5}	10^{-3}	10^{-1}	10 cm

Fig. 1 : Scale of wavelength for the rang of electro magnetic waves.

When a beam of sunlight passes through a prism it is separated into a spectrum of seven easily discernible colour which are :

Colour	Wavelength
Violet	3900-4300 A
Blue	4300-4600 A
Blue-green	4600-5000 A

Green	5000-5700 A
Yellow	5700-5900 A
Orange	5900-6100 A
Red	6100-7000 A

It must be emphasized that the phenomenon of colour is entirely subjective, because it is the sensation created in the brain by the message stimulated by the impact of radiation of a particular wavelength on the nerves in the eye.

The retina of the eye contains innumerable nerve ends which can be differentiated into two categories known, because of their shapes, as the rod and cone types. The rods function when the illumination is of low intensity and do not appear colourless in a dim colour vision. This is demonstrated by the fact that objects appear colourless in a dim light. It is believed that the cones play the predominant part in conveying the colour stimulus to the brain. The exact mechanism is not understood, but it appears that photo-chemical reactions are involved.

Almost any colour, including white (but not black), can be matched by mixing together different proportions of blue-violet, green, and red. These colour are called the additive primaries, and their wavelengths are 700, 546.1 and 435 mμ respectively. It is believed that the eye only responds to these three primary colour stimuli, and all other hues are derived by variations in the relative intensities of the stimuli. The idea was first expressed by *Young* and *Helmholtz*. Who suggested that there were three different kinds of cone-shaped cells responding to the three wavelengths referred to above. This proposed mechanism of colour vision still serves as a good working hypothesis for practical purposes, but is now looked upon as being an over-simplification. Colour triangle is the foundation on which the system of numerical specification of colour based. It is assumed that, the corners of on equilateral triangle are illuminated by the three additive primary colours in such a way that each becomes progressingly weaker until it is completely extinghished when it reaches the side opposite the angle at which it is situated. As seen in the figure, the point at the centre of the traingle is illuminated equally by all the sources and is therefore an equienery white. Every conceivable mixture of the primary additives must exist at some point within the triangle and every colour volve can be defined by describing its location in geometrical and therefore numerical terms.

Materials are coloured because they reflect certain of the wavelengths of the white light which falls upon them, and absorb others. This is referred to as production of colour by the subtractive process. If a beam of white light passes through filters which block the passage of most of the wave lengths and transmit only a selected wave-band, it will be found that three different colours, known as the subtractive primaries, are required to match all hues. These are the familiar yellow, blue and red. It must, however, be emphasized that to match all possible hues, these subtractive primaries must be selected very carefully, and the most perfect combination has proved to be magenta, yellow and cyan. They are in each case white light from which different pairs of the additive primaries have been excluded. The effect of subtrative colour production is shown diagrammatically, where the lights transmitted by partially superimposed filters of yellow, magenta, and cyan are demonstrated.

There are certain attributes of the colour requiring some definition. The hue describes the kind of colour, namely, whether it is a red, an orange, a green, a yellow, or a blue. The saturation of a colour is an expression of the proportion of the dominant wavelength. Thus, ideally, a fully saturated yellow would contain only rays corresponding with a wavelength of 576 mμ. In practice this never happens and saturation is decreased by adulteration with light of other wavelengths. Finally, in order to give a complete description of a colour, the attribute of luminosity must be introduced. This represents the quantity of light reflected or emitted per unit area of the surface. To an observer, colours with the same hue and saturation appear different of their luminosities are not the same.

Defective Colour Vision :

The majority of persons, amounting to 92 per cent in the case of men, are trichromats, meaning that they have normal colour vision. When matching two or more shades with each other they make use of the three primaries. Between 2 and 3 per cent of men are dichromats because their eyes respond to only two primaries and their capacity to discriminate is less acute than that of the trichromate. Finally, there are a small number of monochromats who appear to have cone cells responding to only one stimulus. They are naturally quite unable to distinguish between differences of shade.

It must be borne in mind that about 6 per cent of trichromats have subnormal capacity for colour discrimination and they are sometimes referred to as anomalous trichromats. They can be subdivided into :

(1) Protanopes, whose red receptor does not function.
(2) Deuteranopes, who are not sensitive to green light.
(3) Tritanopes, who are blind to blue radiation. The tritanopes are exceedingly uncommon, but this is not the case with protanopia or deuteranopia.

Testing Colour Perception

The most commonly used device for testing colour perception is the isihara test. This consists of a number of plates composed of numerous spots of different colours and sizes. Incorporated into this random arrangement are a number of dots, tracing out numerals, which are a colour distinguishable from the remainder. The colour sensitivity of the observer's eye is assessed by his ability to read these numbers. It is also possible to detect different kinds of colour blindness with the plates. Four of them, for example, each contain two numbers; one in scarlet and the other in purple. Those who suffer from panatropy cannot see the scarlet spots and only read the purple ones, but the deuteranopes read only the scarlet numerals; those with normal vision see both the numbers.

The Farnswoth test uses 85 pieces of card painted in

different colours and the observer is set the task of arranging them in a logical sequence. The Garner test makes use of samples of material dyed in such a way that some turn greener and others redder in artificial light. To a normal observer any pair will match when viewed in a north light in the Northern Hemisphere, although they reflect different spectral mixtures. To a person who is not equally sensitive over the whole range of the spectrum, however, the two patterns will not appear alike.

There are several studies made on this aspect and many systems developed. The Giles-Archer colour perception unit consists of a lamp with three apertures; one of 5 mm, another 0.5 mm, and finally a 0.5 mm aperture covered with a neutral filter. Filters transmitting six different colours can be placed over the apertures by rotating a disc on which they are mounted. The observer, standing at least 6 ft from the lamp, must name the colours. The observer, standing at least 6 ft from the lamp, must name the colours, using the three apertures in turn. Severe cases of red and green blindness cannot differentiate between the large exposures of these hues, and are often unable to see any of the colours when the small apertures are used.

Description of Ostwald's and Munsell's System of Colour classification

Because of the subjective nature of colour perception and the widespread existence of slight abnormalities of vision a method of classifying an objectively specifying colours is most desirable. Ostwald made one of the earliest endeavours to provide such a system. He specified a series of shades devoid of hue which was in fact, a scale graded in geometrically progressive steps from white of black through greys of increasing depth.

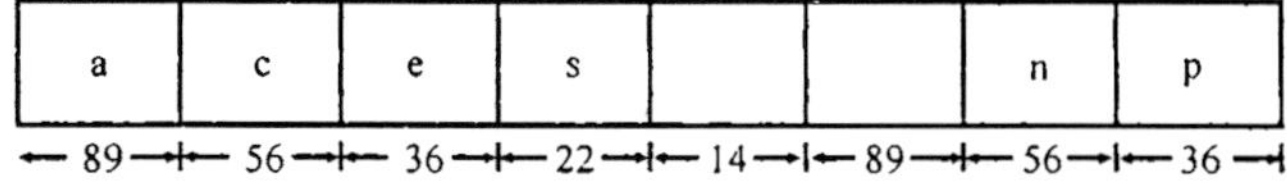

Fig. 2. Neutral Shades in Ostwalds's System.

He also divided the pure hues, in the first place into one hundred colours, but for practical purposes limited the number of twenty-four. Triangular diagrams were constructed for each hue made of the pure colour mixed with varying amounts of black or white, and such a triangle is shown. The parallelogram at the apex marked R contains the pure hue and the parallelograms at W and B are white and black respectively. The row WB contains no hue and is the white, grey, black scale divided into eight graduations. The row RW consists of the hue R mixed with different amounts of white, and the quadrilaterals in RB are made up of the hue with increasing proportions of black.

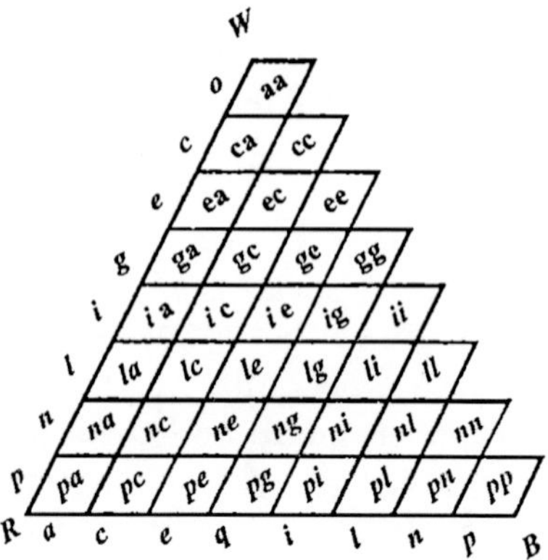

Fig. 3. Ostwald's system of Isochromatic Triangle.

The arrangement of colours shown in figure is known as an isochromatic triangle, and represents the range of shade obtainable by mixing varying amounts of black or white with the hue. Each parallelogram is identified by two letters, the first of which shows the percentage of white and the second the percentage of black mixed with sufficient of the hue to bring the total up to one hundred.

In the case of white, the letter *a* would represent 86 per cent of reflectance, which is the highest obtainable, under normal practical conditions. The scale for the proportions of white might, for example, be as shown below; blacks being the converse.

White (%)	Black (%)
a 86	a 3.6
c 56	c 5.6
e 36	e 8.9
g 22	g 14
i 14	i 22
l 8.9	l 36
n 5.6	n 56
p 3.6	p 89

It follows that *pp* represents the maximum proportion of black and the minimum of white, and *aa* is equivalent to a virtual absence of black. Thus if R were to be hue 24 in the Ostwald scale, square 24, *le* would represent, for example:

White equivalent to	8.9 per cent reflectance
Black equivalent to	8.9 per cent reflectance
Hue equivalent to	82.2 per cent reflectance

And square 24 *ni* would represent :

White equivalent to	5.6 per cent reflectance
Black equivalent to	22 per cent reflectance
Hue equivalent to	72.4 per cent reflectance

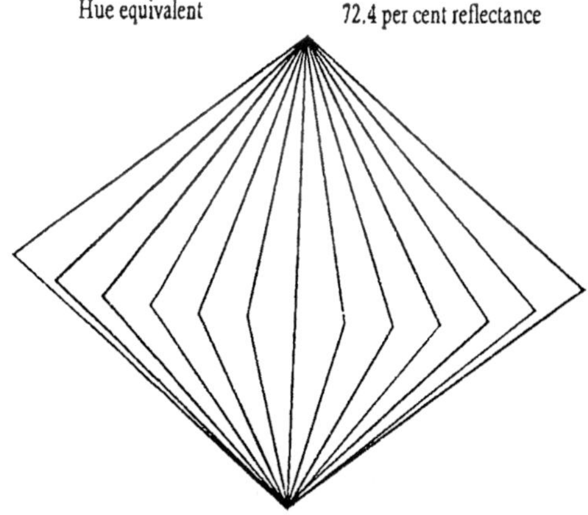

Fig. 4. Ostwald's Colour Solid.

An imitation of a colour solid may be made by mounting the isochromatic triangles, with their black, grey white bases attached, axis to a vertical axis and the apices with the pure hues pointing outwards. (Fig. 4)

The Ostwald classification has been replaced by the Munsell system, which leaves spaces to accommodate pigments of greater brightness than have hitherto been made. The three basic attributes used for the description and location of colours are value, hue and chroma.

(1) Value is the amount of black mixed with the hue. There are ten numbers of the scale, one being a perfect black and ten and white. Neither exists in practice, with the result that numbers two to nine are used.

(2) Hue. There are ten principal hues.

Yellow	Purple-blue
Yellow-red	Blue
Red	Blue-green
Red-purplse	Green
Purple	Green-yellow.

Each of these, however, may be further differentiated into ten subdivisions. As shown in figure, the production are numbered in such a manner that the principal hue is always 5.

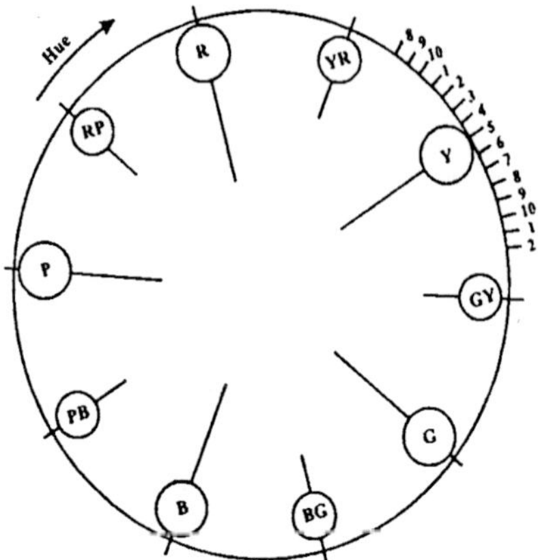

Fig. 5. Munsell System. Arrangement of hues.

(3) Chroma, which is a measure of the saturation of the colour, is expressed by the distance from the value axis. A typical plane in the Munsell chart is shown in Figure 6 in which hue 5Y (yellow) is illustrated in various values and graduations of chroma on the right

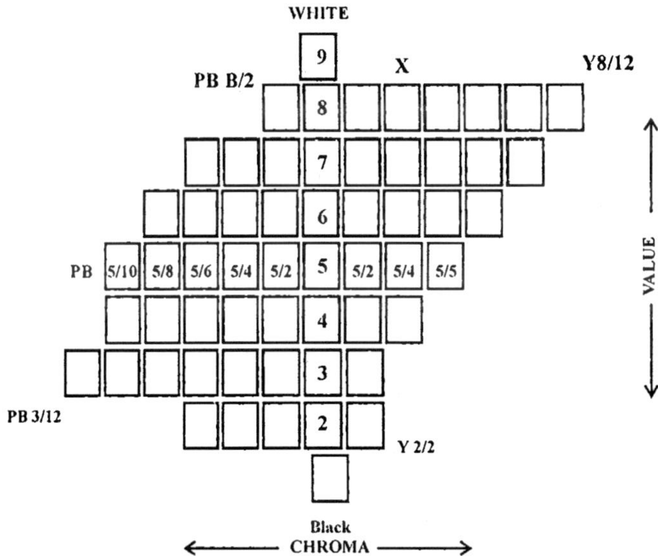

Fig. 6. A vertical section through the-Y-PB plane of the Munsell solid.

of the central black/white axis. On the left of the same axis, the hue 5PB (purple-blue) is treated in the same manner. The individual colours are identified according to their positions and that represented by the square marked X would be designated as 5Y8/4; the hue being followed by the value number and separated by a stroke from the chroma number.

The Munsell chart and Ostwald's system are not all that might be desired because their construction and use not free from subjective elements. They have other limitations such as lack of permanency of the colour standards.

The most convenient method, describing the location

of edour values is to use the co-ordinates, namely the distances at which, the point is situated along lines passing through it and at right angles to the sides of the triangle. In an equilateral triangle, the co-ordinates will always add up to the same total, and it is therefore convenient to treat the sum as unity.

No additive sources of illumination are there that can give the eqvivalents of all the colours in the solar spectrum with equal saturation. Thus, for example , the blue-green of the spectrum must be flattened by the addition of some red before it can be matched with the available primaries. Expressed in another way, the spectral blue-green contains a negetive amount of red in relation to the triangle and is not within the area bounded by it's sides.

Measurement of Colour

Colorimeters usually determine the proportions of the primary additive lights which are necessary to match the colour reflected or transmitted by the sample under investigation. Spectrophotometers differ in that they record the intensity of the reflected light continuously over the whole spectral range.

The lovibend Tintometer is one of the simplest tristimulus colorimeters. This consists essentially of a viewing tube and light reflected from a standard white surface. One half of the field is illuminated by the light reflected by, or transmitted through, the specimen under examination. The other half is illuminated directly by a beam of standard light. Beam coloured glass filters of the subtractive primaries are inserted in to the path of the reflected beam of standard light.and moved by mechanical means.

The coloured glass progress from nearly colourless to fully saturated colours in 200 carefully graded steps. The scales are additive, so that ten units marked 0-01 would equal in colour one glass graded 0.1. The primaries are also such that equal amounts of them inserted in the beam would give a neutral tint. As an example, glass slides, inserted into the path of the beam equal 9.5 units of red, 14 units of yellow, and 4-8 units of blue would transmit a brown colour.

The Lovibond-Schofield. Tintometer is a modification of the simpler form of tintometer in which the readings can be converted into the x, y, and Y coordinates of the C.I.E. (Commission Internationale de Eclairage) system. The instrument is based on the fact that hue and saturation can be matched with two primaries, but the third only affects the luminousity of the colour. Thus in the case of the brown shade described above, which was matched with :

9.5 units of red   4.8 units of blue   14 units of yellow.

Three are 4.8 units of each primary which only contribute neutral tints, and the same effect could be obtained with 4.7 units of red and 9.2 units of yellow if the amount of light passing through the filters could be diminished. In the Lovibond-Schofield instrument the relative illumination falling on the two halves of the field is varied by a movable vane operated by a calibrated control. The illumination is also rigidly controlled as to colour temperature for S_A, S_B and S_C as required. With the aid of monograms the chromaticity coordinates of the operative filters can be determined, and the luminosity is found by the position of the value.

One of the instrument for measurement of colour is the Donaldson Tristimulus Colourimeter. It has been the most successful instrument for measurement of colour but has now been replaced by more sophisticated design.

21

THE TESTING, ANALYSIS AND COSTING OF TEXTILES

The Costing of Textiles

What is Costing?

A knowledge of costing is useful to the student of Textile Engineering, as it enables him to understand the importance of efficiency in the factory and of avoiding waste in the process of manufacture. The study will make him understand how the cost of the cloth and various other articles under his control are closely related. High cost can be reduced and controlled if the factory is well organised, efficiently supervised and waste eliminated. Before we make a specialised study of Textile Costing, it is advisable to introduce to the student the subject of Costing in general.

Cost Accounting

Costing is the collecting, classifying, allocating and controlling of the various expenses in the factory in order to arrive at the cost of producing one unit of a product, process or service as the case may be. The definition will show that costing can be applied in cases of concerns rendering services as well.

The chief objects of maintaining costing are : (1) To find out cost; (2) To discover extravagance and inefficiency; (3)

To eliminate waste, increase production, control expenses and reduce costs.

Estimates

Estimates are not costs. They are only an opinion or predetermined costs with the help of the past experience and records of costing.

Elements of Cost

The cost of product is made up of a number of expenses. These expenses can conveniently be grouped as :—

Examples

(a) Direct Labour	Wages of spinners.
(b) Direct Material	Cost of cotton.
(c) Direct Expenses	Carriage; Royalty for patterns.
(d) Factory Overheads	Oil; Power.
(e) Office and Administration	Salaries of Managers.
(f) Sales and Distribution	Advertising, packing materials.

a + b + c = Prime Cost

a + b + c + d = Work cost

a + b + c + d + e = Office Cost

a + b + c + d + e + f = Total Cost or Cost of Production.

Direct Material comprises all materials, the cost of which can be directly and easily identified with the product.

Direct Labour is the labour spent in the alteration of construction, composition, conformation or condition of the product.

Direct Expenses are expenses which can be identified with or allocated to any particular product. A convenient example is the royalty paid to the authors of books. Royalty paid by Messrs. Binny & Co., for getting the patent rights for using the sanforising process.

Overhead :— This term is used for all indirect expenses, i.e. expenses which cannot be allocated to any particular job but incurred in the course of working the factory for the efficient performance. Overhead expenses can be grouped as Office Overhead and Factory Overhead. The sub-divisions of these Overheads are indirect material, indirect labour and indirect expenses. Other terms for Overheads are Oncost, Burden, Oncharges and indirect expenses.

It must also be noted that where the expenses are constant, i.e. insensitive to the volume of production, they are called "Fixed or Non-variable expenses". Where expenses vary or fluctuate with the output, they are called Floating or Variable expenses. The classification is very essential in view of the fact that while the Fixed Expenses cannot be controlled by the Cost Accountant, the Floating expenses can be effectively controlled to reduce the costs.

Selling and Distribution Expenses :— These are expenses incurred to solicit and to secure orders for the articles manufactured by the factory. Example; Sales managers' salary. The distribution expenses is made up of expenses incurred in the in the actual despatch and packing of goods.

There are various methods of Costing. One essential requirement of Costing is that it must be flexible or elastic and be adaptable to various needs of different industries. A factory cannot be organised to suit the method of Costing. It must be remembered that the principles of Costing remain the same though the methods of application are different.

(1) *Job, Terminal or Contract Costing* : It is applicable in industries where the jobs, contracts are kept separate during the course of manufacture. The unit is the job, contract and each one of them can be identified separately. Example: Building Contracts; Printers.

(2) *Unit Costing* :—It is used in industries where the production is uniform and the units are indentical. Example : Flour Mills; Collieries; Cement Factories.

(3) *Batch Costing* :— Here the orders or jobs are arranged in batches convenient for manufacture. The unit is a batch.

(4) *Process Costing* :— This is employed where (1) production is continuous, (2) the product of one process becomes the material of the subsequent process, (3) the materials change their identify in each process and (4) the

process yield by-product and joint-product. Example: Chemical, textile and soap industries.

(5) *Operative Costing* :— This is a method suitable for the concerns which render service rather than produce articles. Example : Railways, Electricity; Tramways; Banks.

(6) *Multiple Costing* : Time Costing; Target Costs are other methods.

The above methods can be broadly classified as Job Costing or Manufacturing Costing and Operative Costing.

Standard Costing : Standard Costings a widely used term in Costing. It is not a method. It is a system by which the cost is predetermined under normal conditions to compare with actual and thus trace inefficiency and any deviation from standards. It is a yard stick.

Uniform Costing : is a system used as a common method of Costing by different producers of the same industry.

Scrap, Waste :— These are residue materials which arise in the course of manufacture. Where the scrap is used in the process again, it is valued at cost price and the value is oredited to the job/process is which it arose. Where the waste (cotton-waste) is sold, the sale price is credited to the job or process thus reducing the total cost of the process.

N.B. Cost of Excise and other statutory considerations are variable and have to be added where applicable.

Allocation and Recovery of Expenses

As the overhead expenses cannot be charged to any particular job or process or product, they have to be distributed over the entire production. This process of distribution is called Allocation and Recovery. Recovery is the actual charging of the overheads to the products. As a first step the expenses are allocated to the various departments and then they are charged to the products of these departments.

The allocation of these expenses demands a careful handling and planning. The nature of the expenses and that of the industries is required to be studied. An equitable and fair basis is necessary for the proper allocation. The following are some of the common bases of allocation.

(a) Floor area.
(b) Cubic feet occupied.
(c) Actual meter-reading.
(d) Inventory Value.

The recovery can be effected on the basis of :—

(a) a percentage of Direct Material,
(b) a percentage on Direct Labour,
(c) a percentage on Direct Labour-hour,
(d) a percentage on Machine hour.

The first three have simplicity as their merit and require to satisfy certain conditions before they can be applied satisfactorily. The last two have many advantages since many of the expenses depend upon the time.

The Cost Sheet is the statement of showing the cost of production of the product under various heads of cost. The cost sheet can be drawn in any form to suit the particular industry.

The Costing of Textiles

The general idea of Costing gained from the preceding paragraphs should enable the student to follow easily the specialised subject of Textile Costing.

In the Textile Industry the method of Costing will be 'Process Costing.' The factory is supposed to have a perfect organisation and sound store control.

Costing is the proper allocation of expenditure in connection with a manufacturing process so that the cost of manufacturing the finished article can be accurately computed and a price placed on it. In addition it is the responsibility of the costing department to reveal any expenditure or waste and in this way ensure that production is efficient and economical. For the purpose of this we shall consider costing from the point of view of assessing all the items of expenditure and so putting a price on the finished article.

In this lesson we shall consider the detailed costing of cotton yarn and cotton cloth. The figures selected are purely arbitrary and have been selected without regard for current prices. For example, 10d. is selected as the price for a pound of raw cotton simply because this figure simplified calculation.

Costing of Yarn

Apart from the initial cost which includes the cost of the land, buildings, machinery, etc. other items to be considered are as follows :—

(i) Cost of raw material.

(ii) Cost of labour, direct expenses, overhead expenses and office over load.

(iii) Cost of running the machinery.

(iv) Cost of selling.

Note :— The unit of cost in this case is the pound weight, and everything will be determined on this basis.

Raw Material

The first step is to determine how much raw cotton is needed to produce a pound of yarn. Experience normally provides the answer to this question or it may be computed by a test. The percentage of waste depends upon a number of factors, e.g., whether the yarn is carded or combed, the percentage it regains, etc. A typical calculation is given below :

Raw cotton used	100 lb.
Waste	15 lb.
Quantity of spun yarn	85 lb.
5% regain	4.3 lb.
Quantity of yarn produced	89.3 lb.

As our unit of cost is one pound of yarn the calculations is carried as step further.

1 lb. of yarn is product from $\frac{100}{89.3} = 1.12$ lb. of raw cotton.

In the case of combed yarn the same type of calculation is made and an additional allowance is made for comber waste, as follows:

Raw cotton used	100 lb.
Waste (carder)	15 lb.
	85 lb.
Comber waste 20%	17 lb.
Quantity of spun yarn	68 lb.
5% regain	3.4 lb.
Quantity of yarn produced	71.4 lb.

Hence 1 pound of combed yarn is produced from $\frac{100}{71.4} = 1.4$ lb. of raw cotton.

Now that the weight of the raw material required is known the next step is to calculate the value of that weight from the price paid by the manufacturers. Assuming the price of the raw cotton is 10d. then the cost of carded yarn is 10d. × 1.12 which is 11.2d., while on the other hand the cost of combed yarn is 10d. × 1.4 which is 14d. per pound.

Complication arise when the yarn is mixture of combed and carded yarn. However, once the proportions of the mixture are known there is no real difficulty. For example, the cost of a yarn made half from carded cotton and half combed cotton assuming the same price of raw material (10d. per pound) would be calculated as follows :

$$\frac{2\ (10\times1.12)+(10\times1.4)}{4} = 12.6 \text{ pence.}$$

The above are only *gross costs*. They do not represent the true picture but only a part of it. We have assumed the waste constitutes a complete loss. But that is not so. There are two kinds of waste, visible and invisible. 100 pounds of carded cotton give about 13 pounds of visible waste. This may be scutcher droppings, card stirps, card fly as well as other wastes classified as sundries. There is a constant demand for these wastes. In this case let us assume they are sold for 2d. a pound. Thus in 100 pounds there are 13 pounds of waste sold at 2d. per pound, or in other words we recover 26 pence. This means that per pound of carded yarn $\frac{26}{89.3}$ is recovered, i.e., 0.29 pence per pound. This must be deducted from the cost of the cotton. So at this stage the cost per pound may be regarded as 11.2d – 0.29d which is 10.91d.

In the case of combed yarn an allowance must be made for comber waste. This fetches a higher price than carded waste, let us assume in this case 5d. per pound. Thus $\frac{26}{89.3}$

The cost of carded waste is $\frac{26}{71.4} = 0.36$d. per pound.

The cost of combed waste is $\frac{17 \times 5}{71.4} = 1.18$d. per pound.

Note combed wastage 17 pounds per hundred.

Hence 14 – (0.36 + 1.18) = 12.46 pence is the cost of one pound of combed yarn.

It should be noted that while the waste percentage remains constant the price of the waste produced will vary with the price of raw cotton. Hence, whenever the price of cotton changes a new calculation is needed. In practice a table showing these costs for different cotton prices is desirable.

The Cost of Labour (see Figs. 1 and 2)

In a mill which spins only one count the cost of labour may be calculated on the basis of the total wages of the mill for a particular period which is then divided by the total production of that period. In view of fluctuation records would be required to be kept over a fairly long period in order to arrive at a satisfactory figure.

Where more than one count is spun the cost can be

ascertained only by taking the various operations separately. The principle is still the same however, average wages for a period divided by average production.

The wages in the mills are paid to a large extent on piece-work system. So a complete and impartial record of work done is necessary. Departmental clerks are engaged to record the work done by workers. The workers usually have a slip on which details are recorded and the Spinning or Weaving Master in charge signs the slip. The wages are based on the weight of the cotton, spun and woven. The piece-work wage also demands upon the counts. 'Counts' is the technical term for the fineness of the yarn. The higher the Counts the finer the yarn and cloth and therefore superior. The higher Counts demand a skilled hand and therefore the wages increase with the Counts. The width and length are also taken into account in the weaving process, as well as ends and picks etc.

A Register is kept for the workers where particulars of workers and the work done by them are recorded. From this the Pay Roll is prepared at the end of the month. The Pay Roll is classified suitably under the respective departmental heads. Example : In the Spinning Department—Preparation, Roving, Carding etc. In the Weaving Department—Winding, Warping, Sizing Twisting, Weaving, etc. This analysis will help the Cost Accountant to book the Direct Wages to the respective process accounts. Further, this will closely interlock the Financial and Costing Records.

Direct Expenses

The carriage and freight on cotton purchases may be treated as Direct Expenses. The Royalty paid for any particular process or pattern of cloth may be treated as Direct Expenses. The Direct Expenses will be charged to the process to which it is applicable.

Overhead Expenses

The overhead expenses contribute a substantial portion of the cost of a product. The allocation of the overhead and its recovery requires the greatest consideration in costing to obtain the maximum benefits thereof. The different methods of allocation and recovery of overhead have already been discussed. In a textile mill, the overhead is recovered by means of machine-hour rate. The expenses are collected and a rate per loom or spindle is arrived at. The rate is calculated for the cloth manufactured in the loom taking the time factor into consideration.

The overhead expenses are collected by means of 'works order'. Each expense is allotted a number and letter of the alphabet and the expenses are known by this number called works order. Thus Depreciation of Plant will be known as W.O.D7.

These expenses are then summarised and a Departmental Allocation Abstract is prepared. (Fig. 3.) In the case of a factory of average size on such elaborate system is required; it is enough if the financial books are analysed in detail. The basis for Departmental Allocation should be determined after studying the expenses and their incidence.

The Departmental Allocation of overhead will include the following departments.

1. Cotton cleaning
2. Spinning-wrap
3. Spinning-Filling
4. Spooling
5. Warping
6. Slashing
7. Weaving
8. Bleaching, Dyeing etc.
9. Boiler Room
10. Repair Shop
11. Rolling
12. Storing
13. Office
14. Miscellaneous

The expenses allocated to the Service Departments (9, 12, 13 and 14) are in turn allocated to other Production Department on the basis of the service rendered.

Now, the next step is to 'Recover' the expenses. As

already stated a machine hour rate is calculated by dividing the total Departmental Expenses by the number of machines in that department. This rate is applied to the recovery of expenses.

The overheads can also be classified as Fixed and Floating expenses, and the two rates can be used more effectively. In both the cases, the Fixed Machine (Loom or Spindle) hour-rate or floating-machine-hour-rates can be applied.

At the stage owe get the 'Factory Cost' of the cloth or the yarn as the case may be.

Office Overhead

The office and administrative overhead consists of the expenses of office and cost of administration. The office and administration overhead is charged as a percentage on the Works or Factory Cost. In factories, it is included, sometimes in the Works Overhead itself, thus having a 'Composite Rate of Work On Cost'. However, the separation is desirable.

Cost of Running The Machines, Etc.

All costs other than those for raw material, labour and selling come under this heading. Typical items are fuel, lubricant, repairs, transmission materials such as belving, depreciation, etc. This item is a variable one and so accurate records must be maintained. Provided all records are correctly maintained over a period this item can be very easily calculated. Points to bear in mind are that adequate provision should be made for depreciation of machinery. Provision should also be made for fluctuation in the cost of fuel, etc.

Selling and Distribution Overheads

The following expenses will be included under this head :

(a) Store Room for Finished Goods. (Rent etc.)

(b) Travelling Salesmen Salaries, Commissions, T.A. etc.

(c) Sales Manager's salary and Sales Department expenses.

(d) Sales Discount.

(e) Advertising expenses.

(f) Packing Materials, etc.

These expenses should be recovered from the products and the recovery will be made on some local basis. Example:

(a) Travelling Salesmen T.A.D.A. : On the basis of Sales effected in respect of each class of goods or a percentage in unit of production.

(b) Advertising : Actual or Sales Value

(c) Warehousing expenses: On the Works Cost of Goods.

In the case of Carriage and Freight. etc. : These are preferably added to the cost of Invoice price of the goods rather than distributed over the entire production. The goods will be quoted 'Ex godown.' This will help to quote a uniform price to all dealers irrespective of the distance.

Expenses on Displays and Exhibitions are incurred as a part of the selling expenses. These may be regular expenses as a part of sales programme or occasionally they may be incurred in introducting new products or in pushing sales. If they are regular and recurring they can be allocated to the production on the value of sales. If they are occasional, they are treated in the Profit and Loss account. Any special advertising campaign expenses should be treated as Direct Expense. Thus advertising expenses incurred to introduce a new cloth say Sanforised Cloth (by Binny & Co.) should be treated as a Direct expense chargeable to that product only.

With the recovery of this last item the total cost of the product is arrived at.

The selling price can be obtained if the profit is added. The profit is added as percentage on total cost. This percentage will vary. The profit must include the 'return for the Investment', the contingent expenses not included in cost, and a margin for the extra remuneration.

Manufacturing, Trading and Profit and Loss Accounts

A Manufacturing Account is an account drawn up to ascertain the prime cost of a commodity that has been manufactured. It deals therefore only with the raw materials, the charges thereon and direct manufacturing expenses such as productive wages.

In many factories, including Textile Mills, sb-manufacturing accounts are prepared and named after a particular process or stage of manufacture, such as spinning account and waving account.

The principle underlying the preparation of Manufacturing Account can be illustrated as follows :

Let us suppose that the spinning department of a cotton mill purchased cotton worth £8,000. The expenses of the department were as follows :— Freight and Carriage £100. Wages and salaries £1,000. Coal, Repairs £20. Depreciation of machinery £15.

By adding all the above expenses we can find the cost of the spinning department which is £9,610. In other words the cost of the cotton spun in the spinning department is £9,610.

In book-keeping work the relevant information is shown in the form of a ledger account, viz :—

Manufacturing Account
Spinning Department
For the year ending 31st December, 19...

Debit	£	Credit	£
To Cotton	8,000	By trading account	
To Freight and carriage	100	– cost of spun	
To Wages and salaries	1,000	cotton.	9,610
To Coal, oil and fuel	400		
To Gas and water	50		
To Store materials	25		
To Repairs	20		
To Depreciation	15		
	9,610		9,610

Trading Account :— This is an account constructed for the purpose of finding the gross profit.

Gross profit is the excess of selling price over the buying or cost price of the goods sold plus the direct expenses incurred in acquiring them such as Carriages Inward or Importing Cost.

It must be noted that normally a gross profit is made because a trader should be able to sell his goods for more than they cost him to buy or manufacture but a gross loss may arise, if the purchase or manufacturing price exceeds selling price.,

The following example illustrates clearly the simplest form of this account and the principles underlying its preparation :—

Suppose a trader has £9,000 worth stocks of cloth on January 1st; that he purchases during the year cloth worth £1,000; that his sales amount to £2,800 and that on December 31st he has in hand goods which cost him £8,000.

In the first place we must account for £10,000 (the opening stock and cost price of cloth purchased), before any profit arises £8,000 worth cloth is accounted for by the fact that it is still in the traders hand unsold. This leaves £2,000 (10,000 – 8,000) worth goods which have been sold for £2,800 showing a gross profit of £800.

In book-keeping work the relevant information is shown in the form of a ledger account, viz :

Trading Account
For the year ending 31st December 19...

Debit	£	Credit	£
To Stock of cloth		By Sales	2,800
To opening	9,000	By Stock of	
To Purchases	1,000	cloth-closing	8,000
To Profit and Loss			
To Gross Profit	800		
	10,800		10,800

Profit and Loss Account :— This is an account into which all gains and losses are collected in order to ascertain the excess of the gains over losses or vice versa. If gains exceed the losses, the excess is called not profit; if the losses are greater than the gains, the difference is called the nett loss.

Thus in the above example the trader has made a gross profit of £800, but let us suppose that he pays £200 rent, £100 salaries and his other expenses amount to £200. Then his nett profit is only £300 (6800 – £500).

In book-keeping the above information will be shown in Profit and Loss account as follows :—

Profit and Loss Account
For the year ending 31st December 19..

Debit	£	Credit	£
To rent	200	By trading a/c	
Salaries	100	– Gross Profit	800
Expenses	200		
Capital – Nett profit	300		
	800		800

In a textile mill "Manufacturing and Trading Account" may be prepared for each department so that the management can find out the gross profit or loss of each department. The business taken as a whole might be yielding a nett profit and yet one particular depart-ment might be running at a loss. The departmental Manufacturing and Trading accounts will reveal this loss.

Manufacturing and Trading Accounts are also helpful in preparing comparative results of different years.

Export

Special costing calculations are essential for material meant for export as this will involve special additional charges. Besides the prices have to be quoted in a foreign currency. An import duty upon cotton yarns and cloth is prevalent in most countries, which has got to be paid by the exporter to the foreign customs authority. Besides, he has to pay freight, insurance, and packing charges, etc., naturally, he has got to compensate himself in the price when he sells to his clients.

The manufacturer, who carries both home and foreign trade must have his books of costing separately. Otherwise there will be errors in the home-trade costing. The home-trade materials have nothing to do with the export charges, which are meant only for goods to be exported. The costing for foreign trade, in the first instance, is to be done on the basis of home-trades. Thereafter the charges foreign duty, freight, insurance, packing, etc., are to be calculated and added.

Import Duty

If the yarn, is meant for, say Norway, then the Norwegian import duty is to be ascertained first, which will be converted into home-currency and added while costing.

Freight

This item of expenditure will have to be ascertained and the cost per 1b. of yarn added while costing. The charges for the transportation of goods by water is called "Freight". But all other types of transport, if incurred must be included to arrive at the total cost of transport.

Insurance Charges, Etc.

These expenses will include marine insurances charges while the material is on the sea, custom charges and other necessary expenditures, which will be taken into account at the time of costing. Packing boxes used in the home-trade may be sent back to the manufacturers. In foreign trade however, the boxes become the property of the foreign clients and the cost of the boxes should be added while costing. The credit and *discount* also differ in foreign trade from what is prevalent in home-trade. These charges also should be included at the time of costing. All these calculations can be easily kept ready beforehand, one is home currency and the other in foreign currency, in a chart of two columns for ready reference.

Time Study or Motion Study

The data given by Time Study of great help in fixing the rates. As a matter of fact, it is an internal part of costing to-day. No modern management can do without it.

Time study aims at studying scientifically the actual operations of the particular work done by the worker to find out whether any improvements can be made in that working system and also helps in deciding wages.

Time study is a very interesting subject. A person interested in increasing the efficiency of any work must study the subject of "Time Study".

The Costing of Cloth

The costing of cloth of cotton fabrics woven in the grey state consists of three main items which are as follows :—

(1) Materials; (2) Wages and (3) Fixed expenses

Materials

Materials that are required in the weaving of cloth are: warp, weft and size.

Wages

The wages here are those paid for the weaving of cloth, the preparation of yarn for weaving and also the wages of those workers who are paid on the basis of time and not on the basis of piece-rate.

Fixed Expenses

All the expenses for rent, interest, commission, depreciation, insurance, taxes, coal, gas, water, stores, repairs, stationery, law charges etc. are included in this group. For the calculation of the quantities of warp and weft in a particular fabric, the following particulars are required:—

(1) The number of warp threads.
(2) The length of the warp.
(3) Number of weft threads per inch.
(4) Reed width, i.e. space occupied by the threads in the reed.
(5) The length of the fabric.
(6) The counts of yarn.

Allowances for waste should be given while calculating the quantities of yarn for costing requirements. In this connection, it must be remembered that the percentage of waste varies with the class and quality of fabric under process. Generally, in cotton cloth, 5% waste should cover all sorts of waste; allowing 800 yd. in a hank to be used in the actual fabrication of the cloth. The following formula is usually applied to find the weight of wrap and weft required for the cloth, which also includes waste :—

$$\text{(a) Weight of warp} = \frac{\text{Total ends} \times \text{Warp length in yd.}}{800 \times \text{Counts}}$$

$$\text{(b) Weight of weft} = \frac{\text{Picks per inch} \times \text{Reed width in yds} \times \text{length of cloth in inches}}{800 \times \text{Counts}}$$

the cost of sizing materials is to be ascertained separately, then the total cost of size mixing should be divided by the number of pounds that will be sized by that mixing, to get the price per 1b. which may be added to the yarn price or may be treated as a separate item altogether in the costing. An allowances of 5% should be allowed for droppages.

The wages for winding, warping, sizing, drawing-in and weaving are always paid on the basis of piece-work rates which are the accepted standard by both employers and workers. Correct calculation can easily be made for these charges. The costing of cloth can be done in two ways. Firstly from the total number of pieces woven in a particular period and the total cost during the same time, and secondly, by taking each loom into account. The charges for the calendering, folding and labeling also should be calculated while costing cotton cloth.

Contraction and Shrinkage

When calculating the weight for warp and weft due regard should be given to contraction and shrinkage which takes place during the operation of weaving. Contraction takes place in the length of the cloth and shrinkage in the width. These factors vary according to the type of fabric and nature of yarn. Generally for common cotton cloth 6% is allowed for shrinkage and contraction. For example, if the required length of the cloth is to be 50 yards 50 + 6% = 53 yrs should be taken as the length of the warp. Similarly if the width of the cloth to be woven is 25" then 25 + 6% = 26.5" should be the reed width. Besides this shrinkages in width and elongation in length in Bleaching, dyeing, finishing etc. have to be allowed in calculating the amount of warp and weft required. This depends on the varieties of the cloth and the nature of processing and machinery used. Usually about 5% average contraction in width and about 5% average elongation in length is allowed in Bleaching, Dyeing etc. For mercerised cloth an extra shrinkage allowance in width of about 4% (above that for Bleach-ing, dyeing etc.) has to be added. As regards compressive shrinkage in length in Sanforising or Evaset etc., this is based entirely on the actual process require-ments as determined by practical tests and is usually included as a charge on the process.

Dyed and Fancy Cloth

The method of costing for dyed and fancy cloth is just the same as explained above so far as the grey state of the cloth is concerned, but after that stage the items of cost pertaining to the operations the cloth undergoes are add. These operations are bleaching, mercerising, dyeing, printing, etc. So the charges of the finishing processes along with the cost of dyes and chemicals used should be taken into account while costing.

Cloth Contract Terms

(1) A verbal contract is no use but a printed name on a letter may be sufficient.
(2) Any private or supplementary contract is not valid unless in writing, and any variation should be in writing.
(3) Verbal orders on change should be followed by written orders and letters should be copied.
(4) If the seller is in arrears with delivery the buyer may cancel.
(5) If the buyer is in arrears with payment, the seller may refuse to proceed with it.
(6) Acceptance and payment do not close a contract as there may be a liability some time after due to damage.
(7) In cloth contracts definite particulars should be given, and if the particulars are quoted actual, then those particulars must actually be put in the cloth.
(8) If quoted nominal a variation up or down may occur, usually a limit of two counts in yarns used, and half an inch below width.
(9) Keep and mark a sample of cloth and check with the particulars sent.
(10) A fire usually cancels or suspends a contract.

Phrases

The student of costing should also be fully acquainted with the various phrases that are in common use in the industry, such as Forwarding; C.O.D. (cash on delivery); Consignee; F.O.B. (Free on board); F.O.R. (Free on Rail); E. & O.E. (Errors and Omissions excepted); C.I.F. (Cost Insurance Freight); F.O.C. (Free of charge).

Thus if price is quoted F. O. R., it means the price includes the charges of delivering the goods up to the Railway Station.

"Forwarding"

By the term "Fordwarding" is meant the removal of goods from godowns to a railway station for inland transit. The railway charges are not included in the "Forwarding" charges.

All items of cost incurred should be carefully noted to make the costing calculations correct and accurate.

W.No. ...... Shift No. ...... W. No. ..... S. No. ....
Dept : ...... Month .... Dept : ..... Month ......
Machine No. ....

Name : M. Ramachandra

Date		Counts			S.M.	Remarks
	40	60	80	100		
1.						
2.						
3.						
4.						
5.						
6.						
7.						
8.						
9.						
10.						
11.						
12.						
13.						
14.						
15.						
16.						
17.						
18.						
19.						
20.						
21.						
22.						
23.						
24.						
25.						

Counts	Name : Weight	Rate		Amount	
40 60 80 100					
2. 3. 4. 5.	D.A. Rs. ... p.d. ... Medical Fees ...... Overtime. .... Extras				
Deductions :					
1.	Fines				
2.	Recoveries				
3.	P.F.C.				
4.	Miscellaneous				
	Amount Due :				

Piece Work Cards. The day-to-day turn out is recorded on the front side and the wages details are shown by the wages office on the reverse.

Fig. 1.

Production Register

Department :— Counts : Month

Sr. No.	Ticket No.	Name	Dates 1	2	3	4	5	6	7	8	9	10	11
1	751	Ramachandra ..	Sunday							Sunday			

Production Register showing daily output by Counts. The total for day is calculated and Monthly total at the and. Cross checking is done with piece work cards.

Fig. 2.

Summary of Departmental Allocation of Overheads — Works

Month/Period

Allocation Basis	Total	Spinning						Weaving	
Floor Space : Area	30,000	4,500	3,000	3,750	3,750	4,500	6,000	4,500	—
Inventory : Value of Machinery	300,000	—	—	—	—	—	—	—	—
No. of workmen	4,000	600	700	500	800	600	400	400	—
No. of Electric Lamps Production Units									

Sr. No.	W.O. No.	Expenses	Basis	Total Rs.	Spinning Prepara-tion Rs.	Spinning Carding Rs.	Spinning Roving Rs.	Weaving Winding Rs.	Weaving Warping Rs.	Weaving Weaving Rs.	Weaving Sizing Rs.	Other depts.
1.	10	Supervision	No. workers	2,000	300	350	250	400	300	200	200	
2.	20	Depreciation	Inventory	500	100	75	75	75	80	70		
3.	30	Rent	Area	2,000	300	200	250	250	300	100	300	
4	40	Insurance	Inventory	250	50	37.50	37.50	37.50	37.50	40	35	
5	50	Heating	Area	500	75	50	62.50	62.50	75	100	75	
6.												
7.												
8.												
		Service Depts.										
		Power										
		generations										

Summary of Departmental Allocations of Overheads (Works) to Various Departments on the basis shown against each item.

Fig. 3.

Cost Sheet

Month .... Counts ....

Production....... Spinning

	Quantity	Rate per Pound	Amount	Cost per lb.	Cost per Unit Previous Month (Figures not given)
		Rs.	Rs.	Rs.	Rs.
A. Material :—					
Cotton: Opening Stock.	100	2.00	200.00		
Purchase and Issues.	900	2.00	1800.00		
	1000		2000.00		
Less Closing Stock :	150	2.00	–300.00		
	850		1700.00		
Less Wastes.	50	1.00	–50.00		
Cotton consumed. (Total Material)	800		1650.00	2.100	
(1) Direct material :			400.00	–.500	
(2) Direct Wages (Abstract)			1200.00	1.500	
(3) Direct Expenses			400.00	–.500	
		Total	2000.00		
(4) Works Overhead 10% (Total work expenses Rs. 1000)			100.00	–.125	
(5) Office & Adm. Overhead 5% on Factory Cost : (Total office and Admn. expenses Rs. 2000).			100.00	–.125	
(6) Selling & Distribution Cost 20% (Total Rs. 1000)			200.00	–.250	
(7) Total Cost			2400.00	5.100	
(8) Profit 20% (approximate)				1.000	
(9) Selling Price.				6.000	

Cost Sheet for Spinning Department.

The summary shows the cost of spinning yrn.

Note the various elements of cost shown in the Cost Sheet.

Fig. 4.

N. B. Round figures have been assumed to simplify calculations.

Cost Sheet

Month : .....19..

Counts :

Output :
Description : Dhoty
Width & Length : 54 × 8

(1) Materials | (2) Labour

Total Issues.	Per 1b.	Amount Rs.	Process	Total Hrs.	Amount Rs.	Per Unit 1b. Rs.	Per 1b. Previous Month Rs.
			1. Winding.				
			2. Wraping				
			3. Sizing				
			4. Twisting in.				
			5. Weaving				
			6. Bleaching.				

(3) Works Overhead | (4) Summary

Process	Rate per Rs.	Amount Rs.	per 1b. Rs.	Per 1b. Previous month. Rs.	Details	Total Rs.	per 1b. Rs.	Per 1b. previous month Rs.
1. Winding					1. Cost of yarn			
2. Warping					2. Wages.			
3. Sizing					3. Works Overhead			
4. Twisting in					4. Office Oncost %			
					5. Selling and Dist. Cost.			
6. Bleaching etc.								
Total					Cost of Production			

Cost sheet for leaving Department to show the details of cost of production of cloth. Note the work overhead cost and how it is collected and recovered. The works overhead is taken from Departmental Allocation Summary.

Fig. 5

22

METRICATION AND SI UNITS IN TEXTILE TESTING

As the problems of communication between humans deals with language, in the same way textile technologists or researchers face the problems in units of measurement in various elements dealing with this area. A very odd system of linear measurement has been in use by any textile man from quite a long period back in which the basic unit of length is the yard. Commonly in textile, a hank refers to a length unit which is 840 yards used for cotton count testing. And for carrying out a count test, one genenth of a hank is taken which is known as lea. For measuring fabric properties, inch is the unit which is very convenient to use and it One-thirty sixth of the basic yard. In this thread spacing may be indicated as the number of threads per inch. Many times, it necessitates another sub-division of yard to express fabric thickness which may be the thousandth of an inch or the thirty-sixth thousandth of the yard. Many times curions mixture of multiples and sub-multiples of the basic unit is found. All the calculations could be easier, if the multiples and sub-multiples are used in powers of 10, even though the yard remains the basic unit. The basic length unit in the metric system is the meter for calculating long lengths, the long lengths it is convenient to use the kilometer, or the meter × 10 and the millimeter or the meter × 10^{-3} can be used for short lengths.

There is a method of arithomatic converting metric values in to imperial units. But there is always danger of error while converting from one system to another although the method is simple. Besides, the time taken for conversion is wasted. This necessitates the concept of an internationally accepted system. Thus the general conference of weights and measures in 1960 recommended for the universal use of 'Şystem Internationale D'Unit's', which come to be known as SI system. It is a version of the metric system having seven basic units along with two supplementary units for angular measure. The various definitions of this systems are explained as the following. *These basic SI units are* :

Metre : It is the length equal to 1650763.73 wave lengths in vacuum of the radiation corresponding to the transition between the levels $2P_{10}$ and $5d_5$ of the krypton-86 atom.

Kilogram : This is the unit of mass and is equal to the mass of the international prototype of the kilogram.

Second : This refers to the duration of 9192 631 770 periods of the radiation corresponding to the transition between the two hyperfine levels of the ground state of the caesium-133 atom.

Ampere : Ampere is that constant current, which if maintained in two straight parable conductors of infinite length of negligible circular cross-section and placed one metre apart in a vacuum would produce between these conductors a force equal to 2 × 10^{-7} newton per metre of length.

Kelvin : This refers to an unit of thermo-dynamic temperature. It is the fraction 1/173. 16 of the thermodynamic temperature of the triple point of water.

Candela : The candela is the luminous intensity in the perpendicular direction of a surface of 1/600,000 square metre of a black body at the temperature of freezing platinum under a pressure of 101 325 newtons per square metre.

Mole: This refers to the amount of substance which contains as many elementary units as there are atoms in 0.012 kilogram of carbon-12. The elementary unit must be specified and may be an atom, a milecule, an ion, an electron, a photon etc., or a given group of such entities.

Radian : Radian refers to the angle subtended at the centre of a circle by an are of the circle equal in length to the radius of the circle.

Steradian : The solid angle subtended at the centre of a sphere by an area on the surface of the sphere equal in magnitude to the area of a square having sides equal in length to the radius of the sphere.

Table 1 : The Basic SI Unit

Quantity	*Name of Unit*	*Unit Symbol*
Mass	kilogram	kg
Length	metre	m
Time	second	s
Temperature	kelving	K
Electric current	ampere	A
Luminous intensity	candela	cd
Amount of substance	mole	mol

Some quantities used in technology:

Derived units : There are many quantities that are in use in technology. They are defined in terms of the basic units and are referred to as derived units,. As for example —velocity, where the magnitude may be 25 metres per second and the unit in this case being the metre per second. This unit of velocity has no special name but many derived units are given special names . These names are mostly the name of a scientist as a tribute to the contribution to science made by him or her. Some of the examples are the unit of force in the Newton after Sir Isac Newton. The unit of work is the toule and power the rate of working is expressed in watts.

Multiples and sub-multiples : It is already known , to us that, in the kilogram and Milligram, the prefixes kilo and milli means one thousand times and one thousandth of gram respectively. There are other multiplying and dividing values which are given are convenient same times are given in Table 2. The preferred prefixes in the SI system are those which raise the basic unit by powers of 10 in steps of three, as far instance 10^3, 10^6, 10^{-3}, 10^{-6} and so on. It may be difficult to discard the centimeter as a length unit because, it is so convenient practically.

Table 2 : Multiples and Sub-multiples

Units multiplying factor	*Prefix*	*Symbol*
10^{12}	tera	T
10^{9}	giga	G
10^{6}	mega	M
10^{3}	kilo	k
10^{2}	hecto	h*
10^{1}	dec	da*
10^{-1}	deci	d*
10^{-2}	centi	c*
10^{-3}	milli	m
10^{-6}	micro	µ
10^{-9}	nano	n
10^{-12}	pico	P
10^{-15}	femto	f
10^{-18}	atto	a

* The preferred multiplying factors are powers of ten which are multiples of +3 or –3, but there may be special circumstances where the preferred multiplying factors are inconvenient.

The SI System and Coherence :

In the SI systems, all the illustrated units are termed as a 'coherent' set of units. Example-one watt is a rate of working of one joule per one second. Further, the Newton is the force which gives an acceleration of one metre per one second to an unstrained mass of one kilogram. The stress is an the factor 1.0. A derived unit from basic units is itself unity.

Specialised units in textiles :

There are many SI units, which are common to the various sciences and technologies . But a special set of units will be necessary for each special technology to cope with its peculiar problems. This specialized set will have to be acceptable internationally in the same way as the basic SI units. It is because, problems will definition arise for a technologies in France to convert the units used by a technologist in Australia. Table 3 gives a list of SI units recommended for use in the field of textile technology. Hearle in June 1971 has pointed out that, use of multiples and sub-multiples in case of programmes for computers is disadvantageous and all quantities in such cases should be given in basic units which are consistent units. Afterwards, the results may be converted in practical units.

How to convert some basic or specialized Unit in to SI units?

In common practice, many of the references are still in older unit system even when the SI units are in general use. In such cases, a common calculation will be the conversion of data to SI units. The conversion may be carried out from first principles as in the following example.

Example :

Problem : If a terry toweling has an areal density of 8.5 aunces per square yard, what will be the areal density in grams per square metre.

Table 3 : Units in the SI System Recommended for use in Textile Technology (From *Textile Institute and Industry*, July 1973)

Property	*SI Unit*	*Abbreviation*	*To convert to Si Units, multiply value in unit given by actor below*	
Length etc.	millimetre	mm	inch	25.4
	metre	m	yard	0.914
	centimetre	cm	inch	2.54
Width	millimetre	mm	inch	25.4
	centimetre	cm	inch	2.54
Test or gauge length	millimetre	mm	inch	25.4
Thickness	millimetre	mm	inch	25.4
Linear density	tex	tex	Reference should be	
	millitex	mtex	made to BS 947:	
	decitex	dtex	1970	
	kilotex	kiex		
Diameter	micrometre (micron)	um		25.4
	millimetre	mm	inch	25.4
Threads in cloth :				
length	number per centimetre	picks/cm	picks, inch	0.394
width	number per centimetre	ends/cm	ends/inch	0.394

(Contd...)

Warp threads in loom	number per centimetre	endsn/cm	ends/inch	0.394
Stitch length	millimetre	mm	inch	25.4
Courses per unit length	number per centimetre	courses/cm	courses/inch	0.394
Wales per unit length	number per centimetre	wales cm	wales/inch	0.394
Cover factor (woven fabrics)	(threads per centimetre) $\sqrt{\text{tex}} \times 10^{-1}$	(threads/cm) $\sqrt{\text{tex}} \times 10^{-1}$	$\frac{(\text{threads / inch})}{\sqrt{\text{cotton count}}}$	0.957
Mass per unit area	grams per square metre	g/m2	oz/yd2	33.9
Twist	turns per metre	turns/m	turns/inch	39.4
	turns per centimetre	turns/cm	turns/inch	0.394
Twist factor (or multiplier	(turns per centimetre) $\sqrt{\text{tex}}$	(turns/cm) $\sqrt{\text{tex}}$	$\frac{(\text{turns / inch})}{\sqrt{\text{cotton count}}}$ or $\frac{(\text{turns / inch})}{\sqrt{\text{worsted count}}}$	
Breaking load	millinewton	mN	gf	9.81
	newton	N	kgf	9.81
Tearing strength	newton	N	lbf	4.45
Tenacity	millinewtons per tex	mN/tex	gf/den	88.3
Specific stress	millinewtons per tex	mN/tex	gf/den	88.3
Bursting pressure	kilonewtons per square metre	kN/m2	lbf/in2	6.89
Bending rigidity	millinewtons square millimetres	mN mm2	gf mm2	9.81

Conversion Factors for use of Consistent Units

Property	*Consistent SI Units*	*Practicl SI Units*	*To convert to practical units, multiply value in consistent units by**
Length	m	µm	10^6
		mm	1000
		cm	100
Linear density	kg/m	tex	10^6
Twist, thread-spacing, etc.	m^{-1}	turns/m	1
		ends/cm, etc.	0.01
		turns/cm	0.01
Areal density	kg/m2	g/m2	10^{-3}
Cover factor	$kg \backslash/m^{\frac{3}{2}}$	(threads/cm)	
		$\sqrt{tex} \times 10^{-1}$	1
Twist factor	$kg \backslash/m^{\frac{3}{2}}$	(turns/cm) $\sqrt{tex}$	10
Force, etc.	N	mN	1000
Specific stress, tenacity, specific work of rupture, etc.	Pa m^3/kg (equivalent to Nm/kg, J/kg, m^2/s^2)	mN/tex	10^{-3}
		N/tex	10^{-6}

* In order to change an equation from consistent to practical units, multiply each parameter by the reciprocals of these numbers.

Solution : The data in the conversion table states that, to convert aunces to grams we have to multiply by 28.3495 and the factor to change square yard to square meter is 0.8361.

Now the areal density in grams per square meter will be;

Mass in aunces per square yard × 28.3491/0.8361
= 8.5 × 28.3495/0.8361 =
2888.2 g/m2 (using logs)

Another set of conversion tables reduces the labour of calculating the value of 28.3495/0.8361. This type of table provides the conversion factor for changing ounces per square yard to grams per square metre, that 33.9057. Therefore —

8.5 × 33.9057 = 288.198 45 g/m2
= 288.2 g/m2 (to one decimal place).

A further approach is to use a table which converts aunces per square yard directly in to grams per square metre. It is clear from such table that 8507/yd^2 equals 2881.91 g/m^2. And by dividing it by 10 we will get 288.198 or 288.2 g/m^2.

Of course, further publication of special conversion tables for textiles will enable direct conversion of SI units of properties like linear density, tenacity, twist factor, cover factor etc. A number of conversion factors are provided in Table 3. A tenacity of 4.2 gf/den is converted to millinewtons per tex by multiplying by 88.3, that is 370.86 mN/tex.

For textile users some other conversion quide are available in form of slide rules are specially designed. The Shirley metric converter has been provided by the Shirley Institute.

23

APPENDIX - I

Conversion tactors

To convert	Multiply by
Metres into feet	3.2809
Yards into metres	0.9144
Metres into yards	1.0936
Miles into kilometres	1.6093
Kilometres into miles	0.6214
Square inches into square feet	0.00694
Square feet into square inches	144
Square inches into square centimetres	6.4516
Square centimetres into square inches	0.155
Square feet into square metres	0.0929
Square metres into square feet	10.764
Square yards into square metres	0.8361
Square metres into square yards	1.196
Cubic inches into cubic feet	0.00058
Cubic feet into cubic inches	1728
Cubic inches into cubic centimetres	16.387
Cubic centimetres into cubic inches	0.06104
Cubic feet into gallons	6.235

Gallons into cubic feet	0.16037
Cubic inches into gallons	0.0036
Gallons into cubic inches	277
Cubic feet into litres	28.32
Litres into cubic feet	0.0353
Cubic centimetres into pints	0.00176
Pints into cubic centimetres	567.936
Gallons into litres	4.5436
Litres into gallons	0.2202
Grains into ounces	0.00228
Ounces into grains	437.5
Grains into grams	0.0648
Grants into grains	15.4323
Ounces into grams	28.3495
Grams into ounces	0.0353
Pounds into kilograms	0.45359
Kilograms into pounds	2.2046

For the reverse conversion, divide by the constant

APPENDIX - II

Standard regains

Fibre	*Regain (per cent of dry weight)*
Worsted tops, in oil	19
Worsted yarns	18.25
Worsted tops, dry	18.25
Woollen yarns	17
Wool scoured	16
Worsted and woollen cloths	16
Wool noils	14
Viscose rayon	14
Jute	13.75
Shoddy	13
Flax and hemp	12
Silk	11
Mercerized cotton	11
Cotton	8.5
Acetate rayon	6
Nylon	4
"Terylene' polyester fibre	0.5

Counts of yarn

The English count of a yarn is expressed as the number of hanks, of a given length, per lb.

Cotton is calculated on a basis of	840 yards
Spun silk is calculated on a basis of	840 yards
Linen is calculated on a basis of	300 yards
Worsted is calculated on a basis of	560 yards
Woollen (Yorkshire) is calculated on a basis of	256 yards
Woollen (Dewsbury) is calculated on a basis of	16 yards
Woollen (West of England) is calculated on a basis of	320 yards
Woollen (Galashiels) is calculated on a basis of	300 yards (per 24 oz)

Silk, nylon, and rayon ; The denier count is the weight in grams of 9000 metres.

APPENDIX -III

Equivalent British and Metric measures of volume

Gallons		*Litres*	*litres*		Gallons	*Pints*
1	=	4.5460	1	=	0	$1\frac{3}{4}$
2	=	9.0919	2	=	0	$3\frac{1}{2}$
3	=	13.6379	3	=	0	$5\frac{1}{4}$
4	=	18.1839	4	=	0	7
5	=	22.7298	5	=	1	$0\frac{3}{4}$
6	=	27.2758	6	=	1	$2\frac{1}{2}$
7	=	31.8217	7	=	1	$4\frac{1}{4}$
8	=	36.3737	8	=	1	6
9	=	40.9137	9	=	1	$7\frac{3}{4}$
10	=	45.460	10	=	2	$1\frac{1}{2}$
20	=	90.919	11	=	2	$3\frac{1}{4}$
30	=	136.379	12	=	2	5
40	=	181.839	13	=	2	7

50	=	227.298	14	=	3	$0\frac{3}{4}$
60	=	272.758	15	=	3	$2\frac{1}{2}$
70	=	318.217	16	=	3	$4\frac{1}{4}$
80	=	363.737	17	=	3	6
90	=	409.137	18	=	3	$7\frac{3}{4}$
100	=	454.596	19	=	4	$1\frac{1}{2}$
200	=	909.193	20	=	4	$3\frac{1}{4}$
300	=	1363.789	21	=	4	5
400	=	1818.385	22	=	4	$6\frac{3}{4}$
500	=	2272.982	23	=	5	$0\frac{1}{2}$
1000	=	4545.963	24	=	5	$2\frac{1}{4}$
			25	=	5	4
			50	=	11	0
			75	=	16	4
			100	=	22	0

APPENDIX - IV

Comparative temperatures: centigrade and fahrenheit

°C	°F	°C	°F	°C	°F	°C	°F
–5	+23.0	31	87.8	67	152.6	103	217.4
–4	24.8	32	89.6	68	154.4	104	219.2
–3	26.6	33	91.4	69	156.2	105	221.0
–2	28.4	34	93.2	70	158.0	106	222.8
–1	30.2	35	95.0	71	159.8	107	224.6
0	32.0	36	96.8	72	161.6	108	226.4
+1	33.8	37	98.6	73	163.4	109	228.2
2	35.6	38	100.4	74	165.2	110	230.0
3	37.4	39	102.2	75	167.0	111	231.8
4	39.2	40	104.0	76	168.8	112	233.6
5	41.0	41	105.8	77	170.6	113	235.4
6	42.8	42	107.6	78	172.4	114	237.2
7	44.6	43	109.4	79	174.2	115	239.0
8	46.4	44	111.2	80	176.0	116	240.8
9	48.2	45	113.0	81	177.8	117	242.6
10	50.0	46	114.8	82	179.6	118	244.4
11	51.8	47	116.6	83	181.4	119	246.2
12	53.6	48	118.4	84	183.2	120	248.0
13	55.4	49	120.2	85	185.0	121	249.8
14	57.2	50	122.0	86	186.8	122	251.6

15	59.0	51	123.8	87	188.6	123	253.4
16	60.8	52	125.6	88	190.4	124	255.2
17	62.6	53	127.4	89	192.2	125	257.0
18	64.4	54	129.2	90	194.0	126	258.8
19	66.2	55	131.0	91	195.8	127	260.6
20	68.0	56	132.8	92	197.6	128	262.4
21	69.8	57	134.6	93	199.4	129	264.2
22	71.6	58	136.4	94	203.2	130	266.0
23	73.4	59	138.2	95	203.0	131	267.8
24	75.2	60	140.0	96	204.8	132	269.6
25	77.0	61	141.8	97	206.6	133	271.4
26	78.8	62	143.6	98	208.4	134	273.2
27	80.6	63	145.4	99	210.2	135	275.0
28	82.4	64	147.2	100	212.0	136	276.8
29	84.2	65	149.0	101	213.8	137	278.6
30	86.0	66	150.8	102	215.6	138	280.4

Conversion formulae:

Centigrade to fahrenheit : $\times\frac{9}{5}+32$

Fahrenheit to centigrade : $-32\times\frac{5}{9}$

APPENDIX -V

pH intervals over which indicators change colour

Indicator change	*pH range*	*Acid-alkali colour*
Methyl violet	0.1–3.2	yellow-violet
Metaniline yellow	1.2–2.3	red-yellow
Thymolsulphone phthalein	1.2–2.8	red-yellow
Tropaeolin OO	1.3–3.2	red-yellow
Benzopurpurin	1.3–5.0	bluish-violet-orange
Dimethyl yellow (= dimethyalinoazo benzene)	2.9–4.0	red-blue
Methyl orange	3.1–4.4	red-orange-yellow
Tetrabromophenolsulphone phthalein	3.0–4.6	yellow-blue
Congo red	3.0–5.2	bluish-violet-red
Sodium alizarin	3.7–5.2	yellow-violet
Methyl red	4.2–6.3	red-yellow
Lacmoid	4.4–6.4	red-blue
p-Nitrophenol	5.0–7.0	colourless-yellow
Dibromocresolsulphone phthalein	5.2—6.8	yellow-purple
Dibromothymolsulphone phthalein	6.0–7.6	yellow-blue
Neutral red	6.8–8.0	red-yellow

Phenolsulphone phthalein	6.8–8.4	yellow-red
Rosolic acid	6.9–8.0	brown-red
o-Cresolsulphone phthalein	7.2–8.8	yellow-red
Brilliant yellow	7.4–8.5	yellow-reddish-brown
a-Naphsholphthalein	7.3–8.7	pink-blue
Tropaeolin OOO	7.6–8.9	brownish-yellow-pink
Turmeric	7.8–9.2	yellow-reddish-brown
Thymolsulphone	phthalein	8.0–9.6
Phenolphthalein	8.2–10.0	colourless-red
Thymolphthalein	9.3–10.5	colourless-blue
Alizarin yellow	10.1–12.1	yellow-lilac
Tropacolin O brown	11.0–13.0	yellow-orangeish-
Alizarin blue S	11.0–13.0	green-blue

APPENDIX - VI

Specific gravities of ammonium hydroxide at 15°C (*Lunge and Wiernik*)

Specific 15°	*Per cent NH'*	*1 Litre contains g NH'*	*Correction of the specific gravity for + 1°C*	*Specific gravity at 15°*	*Per cent NH_3*	*1 Litre contains g NH'*	*Correction of the specific gravity for + 1°C*
1.000	0.00	0.0	0.00018	0.940	15.63	146.9	0.00039
0.998	0.45	4.5	0.00018	0.938	16.22	152.1	0.00040
0.996	0.91	9.1	0.00019	0.936	16.82	157.4	0.00041
0.994	1.37	13.6	0.00019	0.934	17.42	162.7	0.00041
0.992	1.84	18.2	0.00020	0.932	18.03	168.1	0.00042
0.990	2.31	22.9	0.00020	0.930	18.64	173.4	0.00042
0.988	2.80	27.7	0.00021	0.928	19.25	178.6	0.00043
0.986	3.30	32.5	0.00021	0.926	19.87	1842	0.00044
0.984	3.80	37.4	0.00022	0.924	20.49	189.3	0.00045
0.982	4.30	42.2	0.00022	0.922	21.12	194.7	0.00046
0.980	4.80	47.0	0.00023	0.920	21.75	200.1	0.00047
0.978	5.30	51.8	0.00023	0.918	22.39	205.6	0.00048
0.976	5.80	56.6	0.00024	0.916	23.03	210.9	0.00049
0.974	6.30	61.4	0.00024	0.914	23.68	216.3	0.00050
0.972	6.80	66.1	0.00025	0.912	24.33	221.9	0.00051
0.970	7.31	70.9	0.00025	0.910	24.99	227.4	0.00052
0.968	7.82	75.7	0.00026	0.908	25.65	232.9	0.00053

0.966	8.33	80.5	0.00026	0.906	26.31	238.3	0.00054
0.964	8.84	85.2	0.00027	0.904	26.98	243.9	0.00055
0.962	9.35	89.9	0.00028	0.902	27.65	249.4	0.00056
0.960	9.91	95.1	0.00029	0.900	28.33	255.0	0.00057
0.958	10.47	100.3	0.00030	0.898	29.01	260.5	0.00058
0.956	11.03	105.4	0.00031	0.896	29.69	266.0	0.00059
0.954	11.60	110.7	0.00032	0.894	30.37	271.5	0.00060
0.952	12.17	115.9	0.00033	0.892	31.05	277.0	0.00060
0.950	12.74	121.0	0.00034	0.890	31.75	282.6	0.00061
0.948	13.31	126.2	0.00035	0.888	32.50	288.6	0.00062
0.946	13.88	131.3	0.00036	0.886	33.25	294.6	0.00063
0.944	14.46	136.5	0.00037	0.884	34.10	301.4	0.00064
0.942	15.04	141.7	0.00038	0.882	34.95	308.3	0.00065

APPENDIX - VII

Specific gravity of

Acetic acid at 15°C *Formic acid at 20°C*

Specific gravity	*Per cent*	*Specific gravity*	*Per cent in weight*	*Per cent in volume*
0.9992	0	0.9983	0	0.00
1.0007	1	1.0020	1	0.82
1.0022	2	1.0041	2	1.64
1.0037	3	1.0071	3	2.48
1.0052	4	1.0094	4	3.30
1.0067	5	1.0116	5	4.14
1.0083	6	1.0142	6	4.98
1.0098	7	1.0171	7	5.81
1.0113	8	1.0197	8	6.68
1.0127	9	1.0222	9	7.55
1.0142	10	1.0247	10	8.40
1.0214	15	1.0371	15	12.80
1.0284	20	1.0489	20	17.17
1.0350	25	1.0610	25	21.73
1 0412	30	1.0730	30	26.37
1.0470	35	1.0848	35	31.10

1.0523	40	1.0964	40	35.90
1.0571	45	1.1086	45	40.82
1.0615	50	1.1208	50	45.88
1.0653	55	1.1321	55	51.01
1.0685	60	1.1425	60	56.13
1.0712	65	1.1544	65	61.44
1.0733	70	1.1656	70	66.80
1.0746	75	1.1770	75	72.27
1.0748	80	1.1861	80	77.67
1.0739	85	1.1954	85	83.19
1.0713	90	1.2045	90	88.74
1.0660	95	1.2141	95	94.48
1.0553	100	1.2213	100	100.00

□□□